THE LONELY LIBERTARIAN

Turning Ideas into Gold –
then Gold into Ideas

RON MANNERS

Published in 2019 by Connor Court Publishing Pty Ltd

Connor Court Publishing Pty Ltd
PO Box 7257
Redland Bay QLD 4165
sales@connorcourt.com
Web: www.connorcourtpublishing.com.au

Mannwest Group Pty Ltd
Hayek on Hood
3 / 31 Hood Street
Subiaco (Perth)
Western Australia 6008
Phone: +61 8 9382 1288
Email: mannwest@mannkal.org
Web: www.mannwest.com

National Library of Australia Cataloguing-in-Publication date

The Lonely Libertarian
Turning Ideas into Gold – then Gold into Ideas

Bibliography
Includes Index
ISBN 9781925826579

1. Economics. 2. Australian history. 3. Gold mines, nickel mines & mining – Western Australia: Kalgoorlie region – history. 4. Personal narratives – Australian. 5. Kalgoorlie – history. Manners, Ron (Ronald B.). 6. Banking, investment, economic aspects – Australia. 7. Politics.

Printed in Australia

Cover Photo
A group of Mannkal Foundation's scholars & staff at the Friedman Conference (Sydney) — May 25-27, 2018. Photo by well-known Sydney photographer Josh Rodgers-Falk.

Dedication

This book is dedicated to all the Libertarians out there; even those who don't know much about the word itself.

Libertarians

Those who prefer to make their own responsible life decisions themselves, rather than leaving this responsibility to others.

As Libertarian photographer, Avens O'Brien, says, "I believe that Libertarianism is the most attractive and compassionate philosophy there is. I work to embody that belief and share it with the world." (www.avens.me)

The alternative is Socialism

Socialism and Communism. The two terms were used by Karl Marx, in fact, interchangeably. The Russian Communists still call their domain the Union of Soviet Socialist Republics. Communism is not merely the logical and inevitable end-product of socialism; it is also another name for a socialism that is really complete. We must subscribe, in short, to the definition of Bernard Shaw that "A Communist is nothing but a Socialist with the courage of his convictions."

-- Henry Hazlitt , "The Free Man's Library"

Functioning Values

What is one of the things that I value, above all else? It is the fact that this journey found me travelling with so many truly remarkable people — and that is the 'fortune' that so few people actually find.

Let me also acknowledge those early and more recent adventurers, from many nations, who have brought to our country a set of values that have made Australia one of the few countries in the world to which people flee.

The values they brought with them are diametrically opposed to those of many other countries that have been turned into hell holes—that is, countries from which people flee.

Let us never forget these stark differences when we hear any talk of relativism and the suggestion that all cultures are equal.

How can they be equal when their outcomes are so very different?

Let us celebrate the values of Western Civilisation, property rights and the rule of law that were brought to our shores. Without these values, we might have become just another 'hell hole'.

Yes, we have much to celebrate. And much to be thankful for, as well.

What fools we are,

to miss

the importance

of laughter,

the warmth

of understanding,

and the gift

of friendship,

that wait for us

between the extremes

of love and loneliness.

Nan Witcomb
The Thoughts of Nanushka
Believe in the Dream

CONTENTS

Foreword   - Bill Stacey   9

Acknowledgements   11

Preface   13

Introduction   15

PART I   Starting Out!

1 Why the lonely theme?   19

2 Turning accidents into opportunities   27

3 Why didn't someone tell me (life has no shortcuts)? 37

4 Mannwest's 124-year journey as a family firm   49

5 Northern Australia: the next powerhouse of the global economy?   61

PART II   My 'Fugitive' Years (1975-1982)

6 The alienated Australians   69

7 Adventures in taxation (help feed a starving bureaucrat)   91

PART III   Mining – 'Turning Ideas Into Gold'

8 Triumphs and tragedies of Australia's mining industry: 1960's - 2015
   (The birth of Australia's nickel industry)   165

9 Croesus Mining (Playing a part in Australia's great gold renaissance)   197

PART IV   'Turning Gold Into Ideas'

10 Russia; 'Seven days that shook the world' (Sept. 1990)   209

11 Liberty could be good for you too!   219

Appendix 1   238

Further Appendix *(see www.mannwest.com/the-lonely-libertarian)*

Bibliography   240

Index   241

To Be Really Rich?

For me, measuring things
is more than a hobby.
I feel we can only manage
that which we measure.

How about measuring by the quality of your friends.
These are real values as you travel through life.
Particularly when it comes
to the choice of your wife.

Is it worth the time
and the extra effort?
Even if it can often
bite into our leisure.

So, it's not what we
have in life,
but who we have in our
lives that matters.

So how do we measure
to see if we are rich?
No good measuring in dollars,
that's a unit losing value.

As Mark Twain said;
"No man is a failure,
who has friends."
Without friends – life in tatters.

Money devaluation will drive
dollars down into the ditch.
Far better to measure
with an appreciating asset.

Only one thing
better than friends,
is occasions such as this.
Getting together; a night 'on the piss!'

Ron Manners,
Birthday Poem,
83 on Jan 8, 2019.

Note: 'On the piss' = An Australianism
for a quiet reflective drink with very good
friends.

FOREWORD

To travel is to grow....

It is a long way in distance, culture and perspective from the wide, parched expanse of Australia's 'outback' Kalgoorlie, to the narrow winding paths up the Peak in humid Hong Kong. It takes special insight to love both cities and explain the reasons why that sense has a common cause.

Ron once observed that neither Kalgoorlie nor Hong Kong was a "cry baby city". He elaborated that both were subject to world prices for their goods, be it gold, nickel, elaborate manufactures or human talents. When prices change, these cities have no choice but to adapt, seek new horizons and thereby transform themselves. They don't have the luxury of standing still. They discover and assemble what is needed for success with new prices in a different world.

That unusual transcendence of their economic base comes from values of self-reliance. Survival, dramatic change and progress for both cities don't come from benevolent government, but from adaptation by individuals and their enterprises, despite the impediments of governments and the self-proclaimed wisdom of misguided intellectuals.

The personal escapades of Ron Manners exemplify these city's tales of success against the odds. As Don Boudreaux wrote in the introduction to Heroic Misadventures, Ron is a "searcher", but more than that he has been an assembler of the skills, connections and resources internationally that he has needed for a personal mission exemplified by Leonard Read's phrase "education for one's own sake".

Those who cherish liberty and self-education will find this work an invaluable resource as they seek to adapt to a changing world where the right for individuals to pursue "anything that's peaceful" seems ever more under threat.

-- Bill Stacey, Inaugural Chairman, Lion Rock Institute,
   Hong Kong, August 2019

ACKNOWLEDGEMENTS

My Executive Assistant, Judy Carroll, has been 'on the job' with this book, just minutes after publisher Anthony Cappello came up with the idea.

The remarkable Libertarian Journalist, Nicola Wright, quickened our pace and performed a miracle in getting us 'over the line' with the publishing deadline.

And, always supportive, my patient wife, Jenny, who has graciously endured having several tonnes of paper archives spread from one end of our home to the other.

Chris Ulyatt, has also guided us with his wisdom and suggestions.

Finally, to the rest of our team and many colleagues who have made this journey so enjoyable. If I have missed any of you there is still time as the alphabetically sorted appendix at www.mannwest. com will be a continuing project with monthly updates promised well into the future.

And, thanks to economist, Murray Rothbard, who identified what happens when democracies degenerate into 'mob rule' — "It is easy to be conspicuously 'compassionate' if others are being forced to pay the cost."

Ron announcing the winner of the Australasian Gold Panning Championships at Ballarat, November 2007

Family gathering (including some of Ron's 11 grandchildren) during 'grave-restoration project' for the Swiss, Tamo Great Grandparents, Ballarat – Daylesford, November 2007.

PREFACE

At the time of completing my 2009 *Heroic Misadventures* book (*Australia: Four Decades – Full Circle) 1970 – 2009,* I asked, in despair, how Australia could emerge from such a remarkable growth period and yet be in a state of severe debt dependency.

Now, in preparing my 2019 book ten years later, I look on in similar despair at the lack of leadership, at many levels of politics and business. How many of them can string a few words together in defence of Western Civilisation, that has provided us with opportunities that are the envy of the free world?

Due to an intolerable politically correct burden on our ability to speak with clarity, we see our leaders constantly 'bending over'; and rarely taking a stand or bulldozing their way through the latest politically correct fad that is impeding success and dividing our nation.

Their total inability to appreciate our country and the opportunities it offers will be viewed with alarm by future generations.

Australia has historically been a relatively free society allowing individuals room to manage their own affairs within stable and common law that avoids favouritism and protects life, liberty and property.

As anyone may observe, these virtues are practised most in countries to which refuges flee and least in countries from which they escape.

Australia should welcome newcomers who reinforce our culture and make a clear statement to others, "please feel free to despise Australia, but please do it at a great distance."

The relative success of governments that trespass only lightly on their citizens' liberty is evident and supported by theory. The Austrian economists, especially F.A. Hayek, have demonstrated, to my satisfaction, that even the most benign governments could never assemble the information or command the administrative machinery to serve the diverse and ever-changing interests and aptitudes of the millions of their people. Governments too often exceed their legitimate roles, preventing people from planning and managing their own lives.

Further, experience tells us that few governments remain consistently benign. Lord Acton once observed that power tends to corrupt and the public choice theorists have explained how concentrated vested interests divert democratic governments' policies from the general interest, to those focused special interests, at the expense of people with less ability to organise. Nevertheless, people respond to the good or bad arguments of good or bad opinion leaders and it is of the nature of democracy that their governments follow. Good or at least better government is not a utopian dream but will be achieved only by constant questioning from alert and vigilant citizens.

Without constant pruning, governments will continue to grow. We have allowed governments to grow to such an extent that an appropriate form of pruning may be to start hacking at the roots.

How will we know when we have managed to shrink government to just about right? That will be, the day when the government calls a press-conference, and no-one turns up!

INTRODUCTION

Why another book?

Anthony Cappello (Connor Court Publishing) suggested I write a book about myself.

My immediate response was, "I've already done that, and it's called *Heroic Misadventures*." Anthony laughingly said, "yes, a great book, but that book wasn't about you at all. You were really telling us all about the incredible bunch of people with whom you have travelled your great life. I want you to write a book about yourself this time."

Anthony's invitation was issued around the same time as I received two invitations to give speeches – one a graduation speech to the Engineering Institute of Technology and the other to the Mont Pelerin Society General Meeting in Gran Canaria, Spain. Both invitations gave me an opportunity to reflect on some early influences and how they have led to an ongoing and interesting life.

There were seven 'lost years' which I will refer to as my 'fugitive years' (1975–1982). In my previous book I did touch on a few 'heroic misadventures' that occupied some of those seven years, but I will take this opportunity to delve deeper into my ongoing trials and tribulations with the Australian Tax Department (now called the Australian Taxation Office – ATO).

Having negotiated a satisfactory outcome, I put all this behind me, re-entered the Australian workforce and returned to a second phase of my 'life in mining'.

Life has not always been serious over the past 83 years and the prospect of opening the floodgates of memories made me smile. I happily accepted Anthony's invitation to write my sixth book and thank him for insisting on a tight timeframe.

I hope you enjoy this quick skip through some of the most interesting years in Australia's history.

There will be occasions of minor story-overlap as each chapter stands alone and provides the appropriate context.

A minor confession

Please don't tell Anthony, but I found it extremely uncomfortable writing about myself. However, once I skipped past the initial 'soppy bits' I was much more comfortable – again dealing with the remarkable people with whom I have been so fortunate to have 'travelled'.

I often think that some ordinary things are made extraordinary simply by doing them with the right people. As Mark Twain said, "no man is a failure, who has friends".

It's a simple story

Born to a kind and generous middle-class family in Kalgoorlie, an Australian outback gold-mining town.

I became interested in economics at age 16.

A serious car accident at age 17 put my education on-hold. Nine operations and three years later my education restarted. I have been in catch-up mode ever since.

Our family business, continually operating since 1895 (see chapter

4) has been the launch-pad for many 'heroic misadventures' that have taken me on numerous adventures and through many 'booms and busts'.

These experiences have given me an understanding of money and its usefulness. "Money is a real friend if you can put it to work usefully."

It all sounds very simple but there were surprises lurking around every corner.

...

The Paper Boy Era
1945 - 1947

The Arm-Wrestling for a good cause Era
with Prime Minister John Howard - 1998

The Recovery Era
1953 - 1955

1

WHY THE 'LONELY' THEME?

The theme of the book? It's about 'loneliness'.

What do I know about loneliness, you ask?

In Australia, today, anyone who tries to figure things out for themselves, rather than just go along with the mob, is guaranteed to spend most of their lives being lonely.

However, really it is not all that bad.

Better to be lonely, than to be wrong. Going along with the mob, just for the sake of being popular, often means being surrounded by people without the ability to reason clearly.

The difference between being correct and lonely versus being wrong and popular could be interpreted in this Ayn Rand quotation: "Achieving life is not the equivalent of avoiding death".

Loneliness is a weird word, meaning something different to each of us.

You can be lonely but not alone.

Loneliness is a word often used by victims. Aloneness is for those who are consciously aware! A state in which to experience and enjoy.

It takes courage to be comfortably lonely. Visiting Canadian psychologist, Jordan Peterson, has perfected a way of crafting a message to many of us who feel lonely and uncomfortable with today's 'conformity' demands. By relating to the single relationship, between a husband and wife, he allows us to peer into the current global disorder. By encouraging us to 'sort ourselves out' he encourages us to assume ownership for our own disarray.

His core message of personal responsibility, once taken for granted in the West as necessary for growing up, is now so unfamiliar as to sound revolutionary.

That is how far that we have travelled down this obsessive Politically Correct road of victimhood, toward the 'snow-flake' culture which some people refer to as the 'grievance industry'.

A decaying civilisation lies at the end of this road. Whilst this continuing decay makes critics of Western Civilisation happy, it also stalls the restoration of our productive civilisation.

History shows repeatedly that bad things happen when good people do nothing. In Australia there is no shortage of good people and I'm encouraged by the stirring of so many good people, all seeking the company of other similarly concerned citizens.

––––––––––––––––––

One of the reasons for me to write this book is that I'm constantly asked, "What the hell were you doing for those seven 'fugitive years' and how did you make peace with the ATO before settling back into the role of being a happy little taxpayer?"

For the first few months of my 'fugitive years' I made the amazing discovery that there were battalions of Australian businesspeople who were also fleeing from the level of persecution existing in Australia at that time (the late 1970s).

I rounded up 27 such people for interviews and produced a lengthy document called 'The Alienated Australians' (an edited version appears in chapter 6). These Alienated Australians were from entirely different backgrounds, but all had similar stories.

About the time I left Australia the *New York Times* had sent a journalist to Australia to write a feature article which was titled "The change in Australia's attitude to the work ethic". That article summarised Australian attitudes, at that time. Several interviewees said this article put into words their reasons for leaving. I note a similar 'rising tide' of alienation, in Australia in 2019, as there is again some concern that it is 'easier to vote for a living, rather than work for a living'.

Not so lonely now

In the following pages there will be mentions and some highlights of the 22 years of the Mannkal Economic Education Foundation and how we enjoy the adventure stories of returning Mannkal scholars. Their trials equip them to be independent thinkers, self-sufficient individualists who are able to shoulder responsibilities.

Each of them reminds me of one of my favourite quotations from Ludwig von Mises, the economist:

> Society lives and acts only in individuals; it is nothing more than a certain attitude on their part. Everyone carries a part of society on his shoulders; no one is relieved of his share of responsibility by others.

> And no one can find a safe way out for himself if society is sweeping towards destruction. Therefore, everyone, in his own interest, must thrust himself vigorously into the intellectual battle. None can stand aside with unconcern; the interests of everyone hang on the result. Whether he chooses or not, every man is drawn into the great historical struggle, the decisive battle into which our epoch has plunged us.

With over 1,500 returning Mannkal scholars moving up the various corporate and administrative ladders, I don't feel quite as lonely now.

Mining has been my life

Wherever I go, I'm always asked questions about mining. I'm happy to direct you to the mining chapters of this book where I hope to convince you that mining is the most creative vocation on this planet.

Mining, however, as with most careers in Australia, does have its occasional frustrations.

Let me give you one quick example. It relates to a mining prospect called The Polar Bear Peninsular, near Norseman. I originally pegged that ground in 1979 (yes, 40 years ago) and it has passed through the hands of many companies since then. While I no longer have a financial interest in the project, I follow its progress in the hope that all the ounces of gold contained in the ground may one day be mined and wealth created for those concerned and for the general community.

About a year ago another prominent listed company paid $9 million to acquire that ground and I thought that at last something was going to happen because that company has a nearby mill. So just out of curiosity I recently called the company's Chief Executive. I asked him what progress they were making.

He nearly broke down in tears saying that although they have paid the $9 million and held the ground for a year, they have been unable to advance the project even one centimetre closer to production. He didn't give me any more details. Probably somebody has discovered a nine-legged cockroach or something similar and created some more barnacles on the backside of progress.

A surprising chapter about Russia

In that chapter I will tell you about one of the most interesting experiences of my life. That was to be selected and sent to Russia with a 40-person team for the CATO Institute's *Transition to Freedom* program in September 1990. We had the task of explaining and training Russians on how to handle free enterprise – just at the time they were toppling statues and burning flags as the Soviet Union was collapsing.

It has taken me 30 years to finalise my report on our success or otherwise, but I've just received some statistics that make sense at last. The results contain some interesting lessons for Australia, too!

Good news all the way

How fortunate are we to be living at this inflection point as we move into the 2020s? Let's move upbeat and dwell on this good news of continuing digital disruption in much the same way that the anti-competitive taxi cartel has been utterly smashed by Uber. We will see a similar disruption taking place, but it will be far bigger. It will be nations and countries themselves who will be disrupted.

If I can make one bold prediction now (yes, you can write it down), it is this:

The next big 'industry' to face digital disruption will be our various nations and that children born today will grow up with a radically different understanding of how governments should serve them.

Governments will be forced back to their legitimate role of protection of property rights because property rights, at all levels of society, are vital to a prosperous society and free trade is our future.

People flee **to** countries with property rights and free trade and they flee **from** countries where these qualities are absent.

It is in this new world of national and state disruptions that many of our young graduates will also become 'digital nomads' and the good news is that there are some new tools on the horizon that will make life easier for these 'digital nomads'.

There are many current programs to develop 'digital tools for liberty', including one funded by Mannkal and developed by RMIT, so follow details on this at Mannkal.org.

It is already a new world out there.

There are stark comparisons between how it is becoming harder to do business in Australia whilst other countries are making it easier. That's what our many returning Mannkal scholars tell me and that's what I'm finding from my own travels.

An example: Have any of you opened a bank account recently? You have to take half a day off from work and sit in a bank doing a 100-point test. This is not because you want to **borrow** money, this is to **deposit money**.

Let me compare this with the tiny Nordic country of Estonia. I have e-residency of Estonia. I have an e-residency card:

By plugging it into my laptop I'm in Estonia and, if I chose to do so, I can register a new company. It takes only 10 minutes, at a total cost of 190 Euros – all this, fully on-line and fully in English. While you are at it, you can also open a bank account. That takes another five minutes as they already have sufficient information.

I'm not telling you this just to be cute. I'm not telling you this because it is a tax scheme, because it is not. Taxes have to be paid, but you can now have some choice about where they are paid.

I'm telling you this because, so few Australians realise that the world is passing us by. Our returning Mannkal scholars all have their own stories and they are worth listening to.

It is a new world into which they embark on their careers.

Back to the loneliness theme

Now, if it is still not clear how this feeling of loneliness keeps cropping up, let me quote something that I wrote in 1977, during my 'fugitive years':

> Others will recognise the benefits of 'internationalising' themselves. There is no such country that is simultaneously the best country in which to live, invest and work. The solution is to select a different country for each of the three functions of living, investing and working.

For those of us who seek an improvement in the 'human condition' there is a tremendous challenge to pursue our own chosen plan of action.

The fact that we are only individuals should not deter us.

Solutions won't come from the masses, for the simple reason that the individual is, and always will be, more important than the mass, just as the soloist is more important than the chorus. There may be fine voices in the chorus, worthy of being soloists themselves, but they are lost in the mass.

The mass is merely an agglomeration of individuals, without the possibility of being heard individually. Whereas the solitary lone person is capable of thought and self-expression.

That is why libertarians often feel 'lonely'!

No idea, good or bad, has ever emanated from the minds of a multitude. Ideas come to one mind at a time.

People, it has been well said, go mad in herds, while they only recover their senses slowly and one by one.

The challenge to us, as individuals, is vital for our very wellbeing. If economic sanity can be restored, we can look forward to the benefits and pride that will come from living in a great country with unmatched potential for enterprise and individual initiative.

By working toward a reduction in the size and power of governments we are maximising the freedom of the individual and our creative capacity.

But seriously, only by maximising our freedom to produce, trade and exchange, will we reach our true potential as individuals working for our own individual goals. Our individual goals are sometimes personal profit, sometimes love of a challenge, sometimes love of our country and the freedom one could enjoy in it, sometimes simply the satisfaction of seeing human beings released from poverty.

Professor Ian Plimer demonstrated to students 'how the earth was formed'.

The remarkable Professor Ian Plimer phoned me in mid-winter 2011 and suggested that I take a week off and join him and 40 young 16-year-olds from 'under-privileged Adelaide high schools'. He suggested that we should venture to South Australia's 'outback' Arkaroola in the remote Northern Flinders Ranges for some chilling ridge-top mountain climbing and inspirational talks.

I think my reply was, "why would I ever do that?"

I will always remember Ian's reply. "Because, Manners, if you don't do that, there is every chance that some of these young people will end up in jail."

He explained to me that every year he devoted a week of his life to this experience and concluded by saying, "Manners, you have a story to tell and I have 40 young people eager to listen".

2

TURNING ACCIDENTS INTO OPPORTUNITIES[1]

Birthdays and bush living

It's an absolute pleasure to be here in Arkaroola with you guys. I've met about a quarter of you so far and the other three-quarters I'll meet tomorrow and learn from you.

Earlier this year I had a birthday and I turned 75. I think with modern science the way it is, and if I keep myself fit, I'm about half-way. I'm in it for 150 years and want to maximise the usefulness of the remainder of my life because life is just starting to make sense.

I know that I'm not programmed for certain things and I'm certainly not programmed for the emotion of envy because I couldn't imagine designing my life to come out better than it's come out right now. I look back and realise it was shaped by a series of accidental incidents that just happened. I can't look back and say: "I got a great education, or I went to a great school, or I know all these people". Nothing like that happened to me.

---

[1] *This is part of a presentation I gave to students at Arkaroola in the Northern Flinders Ranges, South Australia, July 2011.*

I come from Kalgoorlie. A rough-and-tumble city where we respect each other for who we are. Nobody in those days had any money. However, nobody was poor. Everyone was independent and nobody lived on welfare. They just worked their guts out to support their families. Now, that was a pretty simple way of how life was, but the key to the success (and the happiness) of those people at that time was that they had no debt. They had no mortgages or interest payments. That was the key to it all then.

Now, what did we do as kids around your age? We had no televisions, computers, computer games, iPods or iPads. What the hell did we do? We made things and entertained ourselves. One such thing is this .22 pistol, homemade by me.

As you can see, it is a thing of great beauty and it is one of my treasures left over from my youth. The other was a mortar bomb with a warhead on it (but I didn't bring that with me.)

Blowing up life

So, what we did without TV and other distractions? We shot things and we blew things up. We all had access to explosives. You could buy detonators, fuses and gelignite from the local grocer's shop. It was called Sheeds, on Hannan Street in Kalgoorlie.

That was not a bad way to go because we didn't know any better.

Then, at exactly the age of 16, one of the first accidental things happened to me. I was working after school in my father's mining engineering business, unpacking crates of machinery that had come from America and in the crates was this packing material. No polyurethane, no bubble wrap or anything like that in those days, so they used to crumple up magazines to make sure the machinery would not move around.

As a kid, I pulled these things out and I smoothed them out and I took them home and read them. They were magazines called *The*

Freeman that came out of the Foundation for Economic Education (FEE) in New York and they contained information about individual responsibility, the free market and all this stirring stuff. The words of Thomas Jefferson. All so inspirational that I could feel my curiosity stirring.

So, how did that link to my being invited to New York, 20 years later, to give a talk to the very same organisation that produced that material originally? I'll tell you later.

Another accident

The second accident happened when I was 17 years old. Driving back from Esperance with my parents one night, there was an oncoming, very old, truck and suddenly its lights failed. It had an overhanging load on the side which just happened to tear the side out of the vehicle I was driving. Unfortunately, some of my right arm went along with it and you might say, it got a bit buggered up at that stage.

The reason I'm telling you this is because I remember at the hospital that night, sitting there, just wondering what's going on. The doctor is saying to my father, "Charlie, we have to take Ron's arm off just here because there is no way in the world of saving it". I recall my father questioned the doctor and this was the first time I had seen someone in authority being questioned. In those days you did not think of questioning the family doctor. He was like God. My father said: "Alan, you may be a very good doctor, you may be a friend of mine, but what I'm going to do it take Ron to Perth tomorrow to get a second opinion". So that's what they did, and I reckon that's worked out well.

About a year later, after they had stitched me up a bit, I learnt to play the trumpet with one hand. I loved music and I missed playing the piano, so I made a little bracket to hold the trumpet up with my little finger and thumb so I could play the trumpet with three fingers. While I was in the hospital undergoing my operations, I used to sneak out in my pyjamas to a local phone box and practise my trumpet in the phone box at midnight. I got away with it too! One night, when my doctor was visiting our ward, he asked, "where is young Manners?" He didn't believe them when they said, "he's down the road practicing his trumpet in the phone box".

About a year later I developed a love for the clarinet. My father and I went to the music shop and my father noticed a beautiful-looking clarinet for £48. The retailer looked at it and then at my fingers, which were all bent. "Your fingers won't go where they are supposed to go on the clarinet," he said. My father looked at him and said: "How much extra will we have to pay if you modified the clarinet to fit his fingers?" The guy replied: "Sorry, but I have never been asked that before. However, I think we could do it".

So, every Wednesday night for years, I was in a little jazz band at The Exchange Hotel in Kalgoorlie having the time of my life. I even played clarinet in Kevin 'Bloody' Wilson's first band when his day job was as an electrician at the Kalgoorlie Nickel Smelter. Again, I thought my father was first-class in that he questioned very good advice and I recall thinking that it all worked out well.

Every time I do my shoelaces up, I think: "It's pretty easy with two hands but bloody hard with one hand". Every time I happen to give a pretty girl a real good cuddle I think: "Better with two hands than one!" Then I say, "thanks Dad". I'll never forget what he did for me.

That accident interfered with my school attendance, so I could not finish high school. However, I had passed one subject earlier in the year – music. One subject would not let me matriculate (now called ATAR) which lead to accident experience number three.

The writing on the wall

I enrolled at the Kalgoorlie School of Mines as a way of catching up. However, I had to catch up on subjects like maths, English, physics, geography and history before I could enrol to do any serious subjects. Also, at the School of Mines I wasn't very good at sport because I still had an arm sling. So, it was suggested I become the editor of the Kalgoorlie School of Mines Magazine. This led to my first job as an editor and resulted in my interest in writing.

I extracted some of the information I'd kept from the packing boxes a couple of years before and started using it in the magazine. Well, the world fell in on me – Kalgoorlie was a very Labor-oriented town and its residents were very unionised. Here I was saying you could be successful if you were an independent individual, but this was in a town where you couldn't possibly be successful unless you were a member of a trade union!

As result, I started getting attacked. People were very abusive, and I thought, this isn't acceptable. I wrote a letter to the Foundation for Economic Education in New York advising their President, Mr Leonard E. Read, that I didn't think his ideas were much good because they were getting me into trouble.

He replied, advising that he thought the ideas were alright. He told me that if I was going to have an opinion on anything, I had to put myself into a position where I had done enough research to defend my position. He said that to help me, they could put me on their mailing list, and I'd receive their monthly information. Mr Read also enclosed a couple of publications for me to look at and offered to speak with me personally if I required any further information.

He ended up being my mentor and a friend for life. We kept meeting all over the world. He was an amazing man.

(Accidentally) engineering greatness

At the Kalgoorlie School of Mines, I wanted to do geology, but the director told me mining was finished. He was almost right, too, as there were more mines closing down than opening up in the '60s. Someone suggested I do something useful, like engineering. I loved electrical engineering, so I got to be an electrical engineer and that was useful. So, it's all accidental as we go.

Looking back, I probably would not have become so involved in these ideas of free markets if I hadn't taken on that editorship and been in a position of having to defend myself. That really got me involved in the world of ideas – I call it 'The Wonderful World of Ideas' because your life is more than just being what you do for a living. It is the concept of why 'one thing follows another' that will make your life interesting. It's the 'human action' side of life that's more interesting to me and it was probably about then that I stopped 'blowing things up'. I realised that any idiot can blow things up and destroy things and graffiti things, but the sheer genius is in creating things like works of art or an enterprise like a company. That's where the genius is – the creativity of being in business. This really helped me get started along that path.

The fourth accidental event occurred one morning in 1967 when I went into the Palace Hotel in Kalgoorlie to pick up a visiting engineer who I was taking out to the mines for the day. He was still having breakfast and suggested I sit down as he was not ready to go yet. He introduced me to the other person at his breakfast table who was in Kalgoorlie recruiting young people to apply for the Duke of Edinburgh Study Conference. They wanted a person from the country areas of Western Australia to go with the other four to be chosen from Western Australia.

He said he'd come to Kalgoorlie especially. He'd been to see the Mayor of Kalgoorlie, who was 93 years old, asking if he knew of any bright young people who would be suitable. The mayor had told him he, "did not know any young people". As result this man was going back to Perth that day with a strike rate of zero. He looked at me and said: "Ron, you are about the right age, why don't you enrol?" He gave me the form and asked me to go away and think about it.[2]

Consequently, I took it away, thought about it, signed up and, after three interviews, was selected. To this day I'm still involved with The Duke of Edinburgh Study Conference. I'm on the selection committee and I go to Buckingham Palace; I'm organising a small event for HRH Prince Philip when he comes to Perth in October this year. He's 90 years old now so this might be the last chance we have to have a good party with him.

Meeting my heroes

All these events are small accidental instances. I've managed to turn these accidents into opportunities and isn't that what life's about? It's recognising a misfortune and turning it into something that really is to your benefit.

The real pay-off for me is it has allowed me to meet many of my economic and philosophical heroes who I had read about in so many books. I have become personal friends with so many of them. With my father's influence, it's given me the courage to question authority from time to time. In particular, the Australian taxation department when it misinterpreted its own Tax Act and tried to extract unreasonable amounts from me. I left the country for seven years and had a remarkable career on the run (that was all reported in my book, *Heroic Misadventures*).

In the end I politely wrote to the taxation department and said: "You are not getting any tax from me, so why don't you tear up the files and I'll come back into the Australian workforce". Well, they agreed. So, I came back.

I enrolled, yet again, at the Kalgoorlie School of Mines at about the

---

2   Philip Lynch (later Sir Philip) also had a part to play in this – see Appendix.

age of 50 and did two geology subjects and formed a couple of public companies. They produced 1.25 million ounces of gold and paid 11 dividends to the shareholders. I paid a lot of tax.

I also joined the Mont Pelerin Society, an international group of 500 economists and scholars. We meet each year in a different country. I was selected as part of a group of 40 who were sent to Russia in 1990 as communism and socialism were falling apart. Our task was to instruct them on how free enterprise worked, as it was arriving the following month! It was quite a challenge when, on arriving in Russia, we found that they had none of the 'building blocks' for a civil society.

I was explaining to a group of Russians how to form a public company, i.e., get 500 people to give us their money so that we can go exploring and discover economic resources and then, with responsible management, pay them a dividend from profits. One man came to me later and said: "I don't know anyone in the world I would trust with my money".

That experience made me realise how fortunate we are to live in a civil society where our chosen colleagues can be trusted and relied upon. We call it Western Civilisation and I only wish it was taught in our schools.

Where to now?

I am now pursuing my three full-time careers:

1. Running the family business: Mannwest, a mining consultancy, has been going strong for 116 years. [Now 124 years as at 2019.]

2. Running the Mannkal Economic Education Foundation: Mannkal spreads the free market message. We have 350 young people attending conferences around the world this year. I'm on the Advisory Council for the Atlas Economic Research Foundation in Washington DC where we co-ordinate training for 485 economic think-tanks around the world.

3. Author: I've just finished writing a book, with another on the way.

    To conclude, I want to leave each of you with something of value to take away. I've noticed that schools or parents are not much good at teaching their young about money. There are two things to remember about money:

- Money is not the measure of a person, but without it you can't achieve much.

- If you are smart and squirrel away some of your earnings, start early, you will be able to get money working for you. Much better having money working for you, than you working for money. Get this one right and you can spend more time on your passions.

It has been a pleasure to spend this time with you all at Arkaroola and to be stimulated by you. Keep questioning the authorities politely but with purpose. Ask me about the emerging youth revolt around the world. I wish you all an exciting life adventure.

Pre-electrification, steam driven winders required extensive tree clearing

Post-electrification. Not a log in sight.

3

WHY DIDN'T SOMEONE TELL ME?
(Life has no shortcuts)[3]

From reading about the Engineering Institute of Technology (EIT) and reviewing some of your award applications I'm extremely proud to join you in this important Graduation Ceremony.

My own education journey was something of a rocky road, as I had a severe car accident during my final year of high school, so I only ended up with one subject, music, which I managed to complete before my accident.

I had to complete all the other matriculation subjects a few years later as a mature-aged student at the Kalgoorlie School of Mines. I tried to study geology only to be told by the School Director: "Mining is finished so you should become something useful, like an engineer". So, I became an Electrical Engineer.

He was quite correct at that time.

I spent the next few years busily electrifying many of the steam-driven mine winders around the Kalgoorlie Golden Mile. That was about the time they started closing down the mines and, in many cases, transporting the winders and controls to other parts of the world.

---

3   A Graduation Speech delivered to the Engineering Institute of Technology (EIT), Perth, 23 November 2018

This brings me to tonight's topic, 'Why didn't someone tell me?' I'll reflect on a few things that I discovered, by accident, on the way through.

I want you to think of life as a very long tunnel (or perhaps more appropriate to think of it as one large diameter electrical conduit?). I'm further through that tunnel than the rest of you, so that gives me the luxury of looking back and laughing at so many of my heroic misadventures.

I have had some giant 'stuff-ups' in my life, to the extent that my fourth book was actually called *Heroic Misadventures: Australia Four Decades – Full Circle*. Guess what? The book was a profitable sell-out and is now available as a free e-book. That book explains how I became a tax refugee from the Australian Taxation Office for seven years and based myself overseas, embarking on the growth industries of that time – namely, money laundering, running hotels in Bali and becoming a rock'n'roll impresario – before returning to my mining career at the age of 50. After running several mining companies, I created a public mining company (Croesus Mining NL) and ended up back at the Kalgoorlie School of Mines, this time studying geology.

Today I won't talk about any of those things. Instead let's talk about that long tunnel called 'life' and conclude with my secret of maximum happiness, sprinkling this journey with a few thoughts that you might find of interest for your exciting careers.

Tunnel of Life

This 'tunnel of life' has created an interesting 'timeline' for me and it demonstrates the usefulness of timelines. My first exposure to timelines was as a 15-year-old. My mother was alarmed that I was learning music without being introduced to the ways in which the great composers influenced each other.

From that point on I have become a 'timeline nut'. More recently

I used a similar timeline to show how the great economists either influenced each other or ignored each other and often repeated the same mistakes, generation after generation.

Then, more recently, as I started tossing out the many tonnes of documents that I have managed to accumulate over 82 years, I came across a great heap of diaries and calendars.

I showed this heap of diaries to one of the bright young Mannkal scholars in our office and asked what I could do with them to separate any gems from the daily 'static' that manages to consume so much of the days of our lives.

He quickly came up with the suggestion that I highlight, in two colours

- Major directional life changes.
- Less important changes but ones still regarded as significant.

We could discard all the rest as static. The two highlighted categories could then be loaded into a spreadsheet and expressed in graphical form. That would give me the satisfaction of inventing a self-imposed screening process and focus on just what these changes have meant to me.

I'm about three-quarters of the way through this interesting exercise and it is somewhat deflating to find that the real successes in my life, for which I have given myself full credit, have in most cases just been accidental events that have come about by being in the right place at the right time. This has given me cause to pick up the telephone and thank people for just being there and, in many cases, just making a casual comment such as, "yes, go for it!"

Once you strip out all the static that makes up, probably 95 per cent of our lives, we can really appreciate the polished jewels that emerge as truly meaningful.

One person, (David Hains) once gave me advice which led me to pick up the phone last year and thank him. Without his off-the-cuff comment, "Ron, I wouldn't be in a hurry to do that", I would not have had the firepower to stand up against a very badly-behaved Australian Taxation Office, hold my ground and eventually win.

I phoned him recently with a suggestion that I jump on a plane and come to Melbourne so that we could enjoy lunch together, as it would enable me to thank him personally for this. His response was "Strange that you mention this Ron. I've just published a book in which you are mentioned, and I wish to thank you also for what you did for me". I had loaned him a mining engineer.

What have I learnt from this shuffling, screening and sorting process? I've learnt that when anyone asks for career advice, you think for a little while and find the simplest way of telling them: "Just keep moving, because you never know the real outcome of a chance comment as paths cross".

Words of Wisdom

At some stage you are going to wonder if I have any words of wisdom to sprinkle into tonight's event. That reminds me of being asked to speak at a recent wedding in Canberra for one of our favourite Mannkal scholar couples. I suspect that I was asked to say a few words about marriage just because I have been married twice. The assumption was that I might have twice the regular measure of wisdom. I don't think it works that way.

Instead of some wisdom on married life from me, I passed on six points from a fellow called Lao Tzu. All of you know, of course, that Lao Tzu was a contemporary of Confucius, in China, back in the 6th Century B.C. Confucius, on the one hand, addressed his message to the "collective". That is why he is the favourite philosopher of mainland China. On the other hand, Lao Tzu addressed his message to 'individuals', like us, one at a time. That is why Lao Tzu has been described as the world's first libertarian.

His wisdom was so insightful that, hundreds of years later, it has been suggested that Thomas Jefferson stole many of Lao Tzu's quotations about government and claimed them as his own. Now, 27 centuries later, Lao Tzu offers classic advice for marital bliss. Given our limited time tonight, I'll share with you just six comments from Lao Tzu about marriage. Very wise comments:

1. "Being deeply loved by someone gives you strength while loving someone deeply gives you courage." (Think about that one).
2. "Marriage is three parts love and seven parts forgiveness of sins."
3. "Love is of all passions the strongest for it attacks simultaneously the head, the heart and the senses."
4. Here's one he addressed to all new brides: "Silence is a source of great strength".
5. One which he equally addressed to brides and grooms: "If you look to others for fulfilment you will never be truly fulfilled".
6. "If you are depressed, you are living in the past. If you are anxious, you are living in the future. If you are at peace, you are living in the present".

Well, for tonight, Lao Tzu also has some advice for you all. Here are his six points on management and careers:

1. He who clings to his work will create nothing that endures.
2. Knowing others is intelligence; knowing yourself is true wisdom. Mastering others is strength; mastering yourself is true power.
3. Care about people's approval and you will be their prisoner.
4. The wicked leader is he who the people despise. The good leader is he who the people revere. The great leader is he who the people say, 'We did it ourselves'.
5. If you realise that you have enough, you are truly rich.
6. To lead the people, walk behind them.

On the topic of wisdom, if we skip forward from Lao Tzu some 27 centuries, it brings us to my bookshelves, bulging with every management book imaginable.

If you were to ask me which of those 100-or-so management books actually made a difference, I would unhesitatingly say, "Only these two":

Two Management
Books

Tom Peters' *Liberation Management* exemplifies the difference between American and European management advice. Americans believe that problems have solutions, whereas Europeans believe that solutions have problems.

Peters, heavily influenced by my favourite economist, F.A. Hayek, understands the use of knowledge in society and how economics is a study of human action, rather than a collection of mathematical formulae.

The other book is, Max De Pree's *Leadership Jazz*. You can see from the yellow Post-its how impressed I was with his description of how an ideal company should operate. Strangely enough, I was doing this three years before the book was published. Let me quote from my 1989 seminar notes to our Croesus Mining team:

The Jazz Band Analogy

Exploration and mining have a natural rhythm, hence the analogy with jazz. What we are doing is like playing in a band. Music means different things to different people but most importantly, deep down inside you, you know when it sounds good – when it all comes out 'right'.

We have an audience. The concerned audience, being 5000 shareholders and the casual audience that cheers when we play a good note and boos when we foul-up. Each individual member of the band is a star in their own right, but generally we are useless unless we are all playing in the same key and better still when we are playing the same tune.

Soloists are good value, but they need the rhythm section, the same way that geologists need mining engineers and vice versa, and without the administration and accounting laying the firm foundation of chords, the geologists and engineers wouldn't be able to play the melody.

We must know when to start, harmonise, blend and stop. As band leader I'm useless without every one of you in this band, just as each of you are totally interdependent on each other. You are all star performers and, as band leader, I care about each of you –what you eat, what you think, what you read (remember you ultimately become what you eat, think and read).

I care about Croesus and its further development so that it can accommodate your career plans and challenges. I'm limited in what I can do about it and realistic enough to know that like a band leader I should limit myself to selecting the tunes and the key that matches our range of skills. The tapping foot is setting the vision for the company (with input from fellow directors and yourselves) reviewing budget limits and then giving you the freedom to set your own whip-cracking pace to go forth and produce the goods.

Leadership Jazz

Coincidentally, three years later, De Pree's book, *Leadership Jazz*, was published. He managed to put the words together in a much more concise form:

A jazz band is an expression of servant leadership. The leader of a jazz band has the beautiful opportunity to draw the best out of the other musicians. We have much to learn from jazz-band leaders, for jazz, like leadership, combines the unpredictability of the future with the gifts of individuals.

That's when I reinforced my belief in the 'flatter than flat' personnel structure which I continue with my 123-year-old family company.

So, as we move toward my concluding remarks, let me share with you the concerns of writer Darius Foroux who was recently writing about listing his own 'useful activities'. He was commenting on doing all those small but useful things every day as it adds up to a life that is well lived. A life that matters. The last thing he wanted to experience was to be on his deathbed and realise that there's zero evidence that he ever existed. Foroux commented on a book that he had recently read, *Not Fade Away* by Laurence Shames and Peter Barton. The book was about Peter Barton, the founder of Liberty Media, who was sharing his thoughts about dying from cancer. Foroux wrote:

> It's a very powerful book and it will definitely bring tears to your eyes. In the book, he writes about how he lived his life and how he found his calling. He also went to business school, and this is what he thought of his fellow MBA candidates: 'Bottom line: they were extremely bright people who would never really do anything, would never add much to society, and would leave no legacy behind. I found this terribly sad, in the way that wasted potential is always sad.'

All of us run the risk of wasting our true potential, so we should never miss opportunities as they present themselves to us every single day. Seize such opportunities with enthusiasm and incorporate them into your own BHAGS (as we used to call those Big Hairy Arsed Goals but now, in this politically-correct era, are called Big Hairy Audacious Goals).

Farm Management

Now, as we move even closer to our conclusion, let me take you back to the early 1960s when I took on the challenge of confronting extreme salinity on some parts of our Esperance farm, in particular one 32-acre paddock which didn't contain even one good acre. I had read in an overseas farming magazine of a new species of salt-tolerant seed that had been developed in Turkey. This *Puccinellia* species was said to grow vigorously on salt-affected land and its sturdy stem, when trodden by animals' hooves, created a suitable surface mulch which enabled other plant species to germinate.

I managed to import a limited quantity of this expensive seed from Turkey and the planting was so successful in the paddock that

I was soon harvesting the seed with a Victa lawnmower hanging off the back of my yellow Jeep (World War II vintage).

Every Esperance farmer that planted *Puccinellia* had similar miraculous results. A couple of years later, I was selected, ahead of fierce competition, for the Duke of Edinburgh's Study Conference. I was always mystified as to why I had been selected but, many years later, when I again met one of the selection panel members, the late Senator Dorothy Tangney,[4] I found out why. She remembered me clearly, which led me to ask if she could remember why I had been chosen. She quickly said, "Yes, you were the bright young bugger who decided he wasn't going to be beaten by a bit of salt and so you went out land found a solution. That was the first *Puccinellia* in Australia and we thought this was a good example of problem solving and, as such, should be encouraged."

You never know where your curiosity might lead you! The relevance of this story is that, 56 years later, I was wondering why Western Australia still had a salt problem in our farming areas when this effective remedy is readily available. It was simply curiosity that caused me to follow up this matter in 2006. After some correspondence I visited the WA Department of Agriculture and was called aside by a friendly bureaucrat who explained the simple reason for the lack of uptake of *Puccinellia* as a remedy to our State's salt problem.

Approximately 40 staff were employed in researching our 'salt problem' and any effective remedy was seen to be a serious threat to their livelihoods. Suddenly, I saw a classic example of the economic theory known as 'Public Choice Theory' (the economics of politics). I was introduced to Public Choice Theory by Prof. James Buchanan, a fellow attendee at the CATO Institute's 'Transition to Freedom' seminar held in Russia in September 1990 to train the Russians on how to handle free enterprise, as it was arriving the following month.

Buchanan and his colleague Gordon Tulloch, along with other

---

4   First female senator elected to the Australian Parliament.

academics, had developed Public Choice Theory to explain how 'bad policy' comes to be adopted politically and is so difficult to eradicate. It also explains how quite often the very worst of people rise to the top. In essence it explains how those individuals who receive the concentrated benefits work the hardest to bring about the changes that benefit them and they design such policies so that the costs are dispersed over many thousands, sometimes millions, of victims/taxpayers. Because it only costs taxpayers 'several dollars a day' they are not sufficiently motivated to march in the streets.

Once you have mastered the principles of Public Choice Theory or the Economics of Politics you will see examples of this in every news item and in every submission from vested interests, at all levels, where they all have their own barrows to push. An understanding of Public Choice Theory will explain how politics in most democracies is broken beyond repair. It is always we taxpayers who pick up the costs, but these are so carefully spread that we simply shrug and put up with it.

It surprises me that there appears to be only one university in Australia (Australian National University in Canberra) that actually teaches Public Choice Theory as an accredited subject. Perhaps it is because this theory may explain how some universities actually work, where the concentrated benefits go to the administrators and staff, and the costs are spread so widely over the students and parents. The teaching of this subject may bring attention to this 'less than perfect' business model. I wish that I had been introduced to Public Choice Theory long before 1990, as it would have solved many mysteries and saved me much wasted time.

I'm so enthusiastic about Public Choice Theory, as a problem solving tool, that our Mannkal Foundation is joining forces with the University of Notre Dame in Fremantle and, for the next three years, Mannkal will be sponsoring a fully accredited undergraduate economic unit. We will be flying in world-class experts to lecture on this topic.

I mention this in all seriousness as the addition of an accredited economics unit like this to your own technical qualifications could be just what it takes to give you the edge on all the other CVs that are lined up in our competitive job market.

Now, to my conclusion, in the form of my festive poem:

Happy and Useful

Aristotle, it has been said,
left us a touch of wisdom.

"Happiness is the meaning
and purpose of life,
the whole aim and end
of human existence."

What did he miss,
what things unsaid?

Emerson later filled in the gap;
"The purpose of life is not to be happy,
it is to be useful, to be honourable,
to be compassionate, to have it make a difference,
that you have lived and lived well."

If we sat Aristotle and Emerson under a gum tree,
would an argument emerge?
No, as each would clearly see,
that happiness is merely a by-product of usefulness.

To be useful
Is to be happy.

Another certain secret
to happiness
is to do just
these two things.

Reduce your expectations of
what government can do for you and others
and **increase** your expectations
of what you can do for yourself and others.

Blend these themes together with your own.
Let courage conquer fear,
as we move into 2019.
Another bloody great year!

26/11/2018

Lapland – Northern Sweden, see pages 58-59
Saturday Entertainment with a Difference!

4

MANNWEST'S
124 YEAR JOURNEY
AS A FAMILY FIRM

On 6 July 2018 I participated in a Skype interview with the U.S.-based Dennis Jaffe, as part of his study of Family Firms that have lived longer than 100 years.

He was fascinated to hear of an Australian family firm that is six years older than the Australian flag or the Australian Constitution.

I told him that there were probably lots of Australian family firms in that category but being 'family firms' they didn't see any point of making a fuss about it.

The following timeline — dates and events, are simply a brief summary of the points covered during this interview and match up with the very volatile timeline graph found at the back of the book on pages 238-239.

Generational Chronology:

William Manners (WM)	Ron's Great Grandfather (Prospector who became a Ballarat Gold Mine Operator)
W. G. Manners (WGM)	Ron's Grandfather (Engineer from Ballarat who started the business in Kanowna, near Kalgoorlie, in 1895).
C.B. Manners (CBM)	Ron's father who operated the business from 1924 to 1955.
Ron Manners (RBM)	Ron now running the business from 1955 – present.

Reason for Longevity

What works and what doesn't (review monthly and reallocate assets accordingly).

Evaluate, improve or exit.

Due to Australia's complex employment laws it is important to minimise direct employees and maximise contractors.

RBM's Brief sweep of the 124 years which included some volatile years.

One crisis after another: -

- Office fire in 1924 **(Crisis # 1)** - All engineering plans and Patent Attorney records destroyed). Same year WGM died of cancer.

- Changed the nature of the business, with CBM taking over and quickly transforming to a mining supply company, rather than direct engineering.

- This was fortuitous as there were no more booms during CBM's 30 years at the helm (these included the years of the Great Depression).

This was a difficult, frugal, 30 years and we 'diversified' by farming in Esperance, adding a business in Norseman (midway between Kalgoorlie & Esperance), whilst still operating our mining supply company in Kalgoorlie.

- 1955, when I signed up to study geology at the School of Mines, I was told, 'there was no future for mining' so I was 'diverted' to electrical engineering. This training was extremely useful during those difficult years when I converted many steam-driven mine winders to electric.

- Our company was almost crippled by the 1962 Menzies Credit Squeeze. (That Credit Squeeze put many of Australia's major & family businesses, completely out of business).

(The 1962 credit squeeze could be described as our **Crisis # 2**, following the 1924 office fire.)

- Coinciding with my assuming 100% ownership (on my father's death in Nov. 1966) I was confronted with the fast and furious 'nickel boom'.

- This transformed everything. (All the world's major explorers and mining companies arrived in Kalgoorlie to replicate the success of Western Mining Corporation [WMC]).

- During the nickel boom we created several subsidiaries, one being a finance company to provide hire-purchase (finance) facilities as the country areas had never regained access to finance.

As an example of Public Choice Theory, our finance company (financing hundreds of office coffee machines & materials handling equipment) was closed by government regulations. A perfect example of Public Choice Theory in action, where the larger finance companies moved back into regional Australia and complained about the 'unfair competition from the small operators'. (**Crisis #3**)

- The other major subsidiary was our own exploration company to explore and develop nickel mines – Mannkal Mining Pty Ltd.

- Then, in about 1972, along came **Crisis #4** which was the very abrupt end of the nickel boom (yes, it just stopped). The nickel price collapsed, as the world's supply normalised again. We were heavily involved in exploration and developing new nickel prospects for joint ventures with major international companies. The 48 people, who we employed at that time, started looking at each other and looked at me as it was obvious, we were 'going broke'.

  I told the staff to sit tight and 'hold the fort' while I travelled overseas for 3 months in a search for highly sophisticated mining equipment that enabled very low-grade mines to still operate economically elsewhere.

  I brought back exactly that, and the next 10 years were a remarkable period where, according to the old saying, "the only people making money out of the mining boom were those selling picks and shovels and sophisticated mining equipment."

- One thing led to another and because the underground diesel trucks, we were importing from Northern Sweden (Lapland) were fitted with Volvo diesel engines we had to send a mechanic to Sweden to be trained. Volvo Truck & Car then heard that we had a trained mechanic, so they asked us to be Volvo truck and car distributors, which we did. That led to another couple of car franchises, including VW and Subaru (just in passing, we sold so many Subaru 4WD vehicles that I was awarded a trip to Japan to spend time at the Subaru factory).

- **Crisis #5.** That's about the time I had to stop working physically (or stop 'earning' in Australia) because of my seven-year confrontation with the Australian Tax Office (ATO). Those remarkable years were covered in several chapters in *Heroic Misadventures* when I ran a hotel in Bali and became involved with a merchant bank, Nugan Hand (pioneering money laundering). We played a vital role in the Reagan Administration's Contra saga.

  Kalmin Exploration Ltd & Malga Minerals NL were two public companies that Mannwest created during the 1970s.

Crisis #6.

- A Perth Accountant whose recent death reminded me how lucky he was to have lived that long, in view of his role in creating our **Crisis #6**, 40 years ago.

- One morning the mail arrived at my office and upon opening one letter I was confronted with tragic news. The Corporate Affairs Department had advised that our family company had been 'deregistered' and as a result 'no longer existed'.

- Serious news, indeed, as that company not only owned the business, the building, my car and my home.

- As was, and still is, the general practice the mailing address for 'company notices' was C/- a public accounting firm, who then lodged all official documentation.

- The reason I now received this latest 'official notice' was that the company no longer existed, hence the accounting firm had no role to play.

- With this letter in hand I marched down-town to the Accountant's office and demanded an explanation.

- The Accountant (now the late Accountant) went pale and admitted that he did recall receiving 'several notices' but could not recall if any appropriate action had been taken.

- After a terse, wide-ranging discussion, I left his office with all our company records, handed to me under some duress.

- After a brief perusal of these files I found, perfectly filed, all the 'final notices' and with all this information I then visited my solicitor's office.

- His news was not good, "It's gone! There is nothing you can do." Well, there was something we could do.

- It required top-flight lawyers many months to successfully 'set aside' the Court Order that had been made to 'liquidate' this now defunct (but asset rich) company.

- I resisted the temptation to do physical harm to that Public Accountant, at that time, but did quickly appoint a replacement Accountant. To this day, I regularly ask, "Have all required forms been lodged on time?"

Australia again rescued by gold

- Rescued by gold increasing in price which stirred the sleeping gold industry, 1979 – 1980. There was a price spike in Jan. 1980 due to substantial gold purchases made by Middle-Eastern countries when they were flush with 'oil dollars'. Gold soared from US$38 in 1970 to US$860 in January 1980.

- During the interesting years, between 1982–85, Mannwest developed a series of public mining companies (Coopers Creek NL, Mistral Mines NL, King Mining Ltd, Great Central Mines NL), before creating our own public IPO, Croesus Mining, in 1986.

- The importance of Mr H.M. (Harry) Kitson, (probably Western Australia's most senior and respected Accountant), being on my family company Board, from 1972 to his death in Oct. 2000. He was a steadying figure and key advisor during these extremely difficult years and he also became a director of Croesus Mining.

- 1997 saw the establishment of my own Economic Think Tank – Mannkal Economic Education Foundation (www.mannkal.org).

- The years 1996 – 2019 have been busy years, expanding Croesus Mining before it became associated with Canada's Eldorado Gold, plus expanding Mannwest Group and the Mannkal Foundation.

- This brings us up to the present year, 2019, with my time allocation approx. 33% each apportioned to:

  o Mannwest Group Pty Ltd (Family Company) – www.mannwest.com
  o Mannkal Economic Education Foundation (Think Tank) www.mannkal.org
  o Writing

What putting this business timeline together has told me?

My business life would have been significantly diminished had I not, almost by coincidence, been connected with these individuals:

Sir Arthur Fadden – the inventor of Australia's 'loophole industry'.

Keith Parry – best mining career advice.

Harry Kitson – my business mentor.

David Hains – introduced me to the world out there.

Clyde R Maxwell – Californian tax attorney.

Simon Lee – helped me over a final hurdle.

These individuals all deserve a special place in this book's appendix.

Mission Statement
For mining companies that I have operated:

❖ Create profits and distribute according to the following formula:
 	
    o   One third to shareholders for them to do as they wish.
    o   One third to exploration (the future).
    o   One third to investment reserves (the insurance policy).

For our family company Mannwest Group (operating since 1895):-

❖ Create profits and distribute according to the following formula:
 	
    o   One third to shareholders for them to do as they wish.
    o   One third to future generations of Australians (the future). (*By investing in Mannkal Foundation*)
    o   One third to investment reserves (the insurance policy)

THE GUY AT THE NEXT DESK
(and his four lessons for life)

My year of enforced business training with Noyes Bros[5] (sales engineers) was 1954. I arrived in Melbourne as an 18-year-old country kid. All those around me appeared to have the world under control, but I felt as though I was in a stage play with excellent actors, although I hadn't quite been told the plot.

On my first day the guy sitting next to me seemed reasonable, he was told to "show this new guy around".

I worried about him as he didn't join us all for a pie and beer at lunch time nor after work. He just said, on both occasions, "I have to run".

At the end of the day I asked him, **"well, what's my lesson of the day?" I clearly remember his advice. "There are four things you can't get too much of – vegetables, fish, sex and laughter."** I liked him straight away and we got along well, despite his daily habit of 'running'.

I knew his name was Ron Clarke but didn't realise that a year-or-so later he would be chosen to light the Olympic Flame during the opening ceremony of the 1956 Summer Olympics in Melbourne.

He went on to win nine Australian championships and 12 Victorian track championships before becoming an international running star which resulted in him living in Europe for 14 years. During a 44-day European tour in 1965, he competed 18 times and broke 12 world records, including the 20,000 metre (12.4 miles). On 10 July, at London's White City Stadium, he became the first man to run 3 miles in under 13 minutes, lowering the world record to 12:52.4. Four days later, in Oslo, he lowered his own 10,000 metre world record by 36.2 seconds to 27:39.4, becoming the first man to break the 28-minute barrier.

After he returned to Australia, in 1995, he became the Mayor of the

---

[5] Clough Engineering acquired Noyes Bros in 1991.

Gold Coast (2004–2012). Our contact over the years was spasmodic as we watched each other's careers. We nearly caught up with each other in London when he was running a 'gym-club' but settled for a long phone conversation. My last contact was when he established a philanthropic organisation in Melbourne as we both had serious concerns about why there was so little philanthropy in Australia.[6]

Ron Clarke featured in a Nuova Campioni dello Sport
(New Champions of Sport) 1967/68 Sticker -n. 44.
Wikipedia Commons.

---

[6] See *The Australian*, 18 January 2015.

REINDEER SLAUGHTER
– A LAPPISH CUSTOM

My search for sophisticated mining equipment took me to Northern Sweden. Sixty kilometres from Kiruna, on 16 September 1972, we became the first Australians to witness a reindeer slaughter.

A decision, to slaughter, is never made until the actual day before the slaughter so rarely do even the Swedes hear abut this event and have an opportunity to witness it for themselves.

The main occupation of the Lappish people, in North Sweden, is keeping reindeer herds which provide them with meat as well as a livelihood from selling the meat.

The reindeer are first herded into circular yards where they are separated into groups each having the particular brand of their owner. When grouped, in this fashion, the Lappish herdsmen then select the female reindeers which are going to be slaughtered. These must be killed in the early autumn, before the meat is spoiled by them becoming 'love sick'.

The whole performance was carried out in a rather primitive way by using their traditional techniques of killing the reindeers with a knife which they stick into the heart, then letting the animals run, draining themselves of blood.

The reindeer meat to be sold, outside of the Lappish community, must be treated another way and after slaughtering this meat is distributed throughout Sweden via freezer trucks. These reindeer, for the outside market, are usually shot with the final cutting being done in an abattoir elsewhere.

The reindeer exist as the prime purpose of food for the Lappish people. Over 90% of the animal being used in this way. In the late autumn they have a festival where it is customary to eat 100% of the animals including the marrow from inside the bones. This is extracted with long sticks. They mix the marrow with liver which is cut into fine pieces. Even the reindeer horns are ground up into a fine powder for export to Korea where it is used as an aphrodisiac (Australians are known not to require this).

The Swedes like smoked reindeer meat as well as dried reindeer for chewing (like chewing gum).

It was noticed that several reindeers were neither slaughtered nor released. These were reindeers bearing no mark. They are later sold by auction with all proceeds going into a common Lappish fund to finance their festivals. They are rather wild affairs with copious quantities of homemade aquavit and ether sniffing.

These Lappish people follow their reindeer herds from the mountains to the coastal areas each season, some still living in traditional Lappish huts. These consist of tent-like buildings made from wood with clay filling, with a fireplace in the centre.

Although rather a macabre sight, the traditionally dressed Lappish people walking around with their buckets of blood and pieces of flesh and liver, created a scene not entirely without some natural charm.

Iron ore for the world, from Roy Hill in Australia's North

Enterprise Zones – Examples from around the world.

5

NORTHERN AUSTRALIA: THE NEXT POWERHOUSE OF THE GLOBAL ECONOMY?[7]

Do you believe in miracles?

Let me tell you about one miracle that confronted me in Hong Kong back in 1978. It left me with a lasting impression. Back in the late 1970s our mining engineering firm in Kalgoorlie was the sales and service agent for the pioneering low-profile diesel truck from Sweden called the Kirina truck.

Australia's leading mining company of that time, Western Mining Corporation (WMC), was running a fleet of these trucks in its underground decline shafts. These trucks were the lifeline for its production. Any breakdown meant no production from one of their decline shafts, and every lost hour of production cost our client $2,500 per hour, a huge amount of money in those times.

Most of the time we had these spare parts in stock and we could

---

[7] *Keynote Address to 3rd Annual North Queensland Resources Development Conference in Townsville, Queensland – 30 August 2010*

get them from Kalgoorlie to Kambalda in 22½ minutes flat, but about once per month we had to fly spare parts from Northern Sweden to Kalgoorlie. Now, whatever we did, we couldn't beat 3½ weeks. That was 19 hours on a plane and 3 weeks getting them through Customs in Perth (even though we paid a high price for a Customs Clearance Agent). Well, that's what we called 'the normal' by the standards of excellence that operated at the time. Such breakdowns cost our client $2,500 x 3 x 7 x 24 = $1.26M per 3 weeks.

What's that got to do with the Hong Kong miracle?

Well, at about that time, I was passing through Hong Kong and saw in an optician's window a sign advertising a new invention called contact lenses. So, I thought I'd better have a pair, called in, got myself measured up and asked when they would be ready? They assured me 10 am the next day.

I inquired if they had an on-site technician that enabled them to give such good service. "No", the proprietor responded. "We get them from West Germany overnight and if we get the prescriptions through to Germany on the 'telex' (pre-facsimile and computer days, of course) by 4 pm, they are always here by 10 am the next day." This service, to me, seemed incredible and I said, "But what about getting them through Customs?" With a shrug of the shoulders, he answered "But we don't have any Customs Department to get in the way".

This set me thinking about what role the Customs Department was playing in respect to our mining spare parts, especially when there was no customs duty payable and in no way could they ever be deemed to be a prohibited import.

About the same time, I started reading about the concept of Enterprise Zones being established around the world to cut through the bureaucracy and facilitate commerce. In short it was the birth of the term that's now widely used— 'competitive advantage'—and centres that were establishing these Enterprise Zones, or EZs, were stealing a march on other regions which continued to be smothered by bureaucracy.

At that time, Kalgoorlie (a service centre like Townsville) was going through a tough time. The nickel industry, which had collapsed 10 years earlier, showed no signs of recovery and gold had dropped from US$860 to US$280 per ounce. Kalgoorlie was developing a regional service centre concept and picking up a fair bit of engineering work from the fast-developing Pilbara region to the north.

As President of the Chamber of Commerce, I was concerned that we needed to become an EZ and really put into practice our motto: 'servicing the mining industry from Kalgoorlie'. So, I took three months off and visited several international Enterprise Zones. More importantly, I spent time with the several free-market think-tanks in the US that were actually 'breeding' numerous EZs. In this they had been encouraged by President Ronald Reagan whose idea was to 'flood the world with EZs' as a way of breaking down trade barriers and encouraging international goodwill via free and voluntary trading between nations and regions.

The concept, of course, had been taken from two millennia ago when the Roman Empire used the EZ concept for the Port of Delos, where they managed to out-compete the Port of Rhodes. How EZs evolved into the modern version, where there are now over 3,000 such zones in 135 countries is outlined in the August 2010 edition of the *IPA Review*. The article is titled 'Unleash the North'. It is now quite exciting to see these regions evolve after my earlier involvement in 1982.

EZ in WA

Well, you might ask if I was successful in getting the EZ concept up and running in Western Australia. The answer is 'nearly'. The

State Government became very keen and the relevant Minister for Commerce, Barry MacKinnon, showed sufficient interest to visit some zones and assemble some proposals. Subsequently, the Minister for Regional Development, Julian Grill, commissioned concept proposals to gain support at various State Government levels.

However, the labour unions were most vocal critics of our plans, insisting that it was all about importing 'Coolie Labour' to compete with their labour monopolies. So, the idea died. But, some months later, I received a phone call from the Administrator of the Northern Territory.

"We are interested in establishing an EZ in the Northern Territory and I understand you are an expert on EZs, so can we obtain some information from you?"

So, my EZ archive box was promptly couriered to Darwin. It's interesting to see how WA responded to the news that the Northern Territory had established an EZ. WA didn't want an EZ but were upset when the NT got one!

So, what was the eventual fate of the Northern Territory EZ? It didn't work, mainly because the NT EZ was run by the government, and we all know that the primary goal of government is not 'improving the standard of living of its subjects', it is simply to 'perpetuate and extend the power that it has come to enjoy'.

EZs are about reducing power and regulations, so that's why the NT Zone is not a good example. But that was THEN (when there really wasn't much happening in Australia). NOW we are serving the world and can be far more ambitious about our concept planning.

Every time I step on and off a plane, anywhere in the world, I'm reminded of the significant role that Australia is playing in developing resources around the world. Planes are full of Australian geologists and mining engineers bringing projects to reality in remote parts of the world that will out-compete our own home-grown projects, unless we can shrug off the burden of the myriads of regulations and high taxes.

There is a new competitive world out there that we need to meet head on. And we need to rediscover our 'competitive advantages' that will continue slipping away unless and until we cut through the morass of green tape, red tape and black tape that is throttling so many new projects here.

In March 2010, I attended the Mines & Money Conference in Hong Kong and I saw a stark example of how Australia's competitive advantage was slipping. Of the Australian mining companies represented in the various booths at that conference, simply by asking around, I found that 65 per cent of their exploration budgets were already being spent outside Australia. That was before our Federal Government announced the Anti-Mining Tax, which has seen this percentage of exported exploration capital dramatically increase.

So how have EZs fared in other countries since their modern version appeared on the scene in the 1980s? They have been used extensively in China, specifically for their ability to stimulate business, and they are responsible, in no small way, for the remarkable Chinese economic miracle that we are currently witnessing.

In some US states, EZs have been used to encourage economic growth and investment in especially depressed communities. While the goals may vary, in general, incentives are given to businesses locating within the zone boundaries to invest capital and create jobs. The question is also being asked, "If EZs are good for business, why not apply the same principle to individuals?" In that way, not only will businesses operating within those jurisdictions be free of tax burdens, but so will individuals who choose to live in those jurisdictions.

One of the greatest problems that I see in Australia, at this time, is the extreme difficulty faced not only by companies but by individuals to create a pool of capital with which they can get established as a business or as a family unit. How difficult is it for a young couple to become home-owners without incurring a life-long mortgage commitment? A little-known fact is that Australia's household debt is the highest in the world.

The establishment of such EZs in selected parts of northern Australia would give a tremendous incentive for such young couples to move north for a few years and work in such an environment, thereby enabling them to enjoy an 'unencumbered life'. So, there is a noble objective behind these EZs!

Again, looking to the US model, since 1981, 43 states have implemented EZs, simply because they work. While some of these EZs are more effective than others, some places were screwed up so badly that they were beyond help.

The theory is clear

Fewer costs increase the likelihood that a business will earn a profit and be able to hire more workers. EZs that at least lower the tax burdens imposed by governments may convert an area that is otherwise considered too remote to be competitive into a financially viable region.

There are many questions that can be asked about the potential benefits from creating these EZs. For example, what does the Federal or a State Government stand to lose by exempting them from taxes? Governments may, in fact, be net gainers from such a move: the additional incentive and commercial opportunities may lead many currently unemployed people back into the workforce, no longer draining government coffers with unemployment payments.

How hard will it be for governments to keep track of who should be paying tax and who shouldn't be? With the present computer surveillance on all companies and individuals, all details of our movements and activities are already 'on file'. It's in the context of the general idea and history of EZs detailed above that a group has been formed to study the benefits of EZs in more detail, against our high-cost background.

Unfortunately, the Australian public is generally unaware of the damage done by high taxing, high regulating governments. No surprise when you consider that the government is in charge of gathering and distributing statistics and distributing them to government-licensed media, for distribution through government-controlled schools. These are still the prime sources for public information even though the ABS, Treasury, the Productivity Commission, the think-tanks, etc all have their own websites.

If the public received the uncensored facts, we would have had a public revolt long ago. Quite a few commentators and think tanks are already echoing the words of Thomas Jefferson who once said: "I hold it, that a little rebellion, now and then, is a good thing, and as necessary in the political world as storms are in the physical world".

These words echo the level of exasperation felt by many Australians as they try to go about their regular productive business activities; and this is the key thought I wish to share with you today:

Australia boasts that it has a highly educated population, but if that were true, it would be obvious to all that if you tax something, you get less of it. If you subsidise something, you get more of it.

In Australia, we continue to tax work, growth, investment, employment, savings, productivity, initiative and ability, whilst subsidising non-work, consumption, welfare and debt. No wonder we are getting less of the former and more of the latter!

So, let me tell you how you can join in this peaceful revolt, already joined by over 50 companies, plus their senior executives, in developing this regional zone concept for Northern Australia.

The group behind this plan is known as Australians for Northern Development & Economic Vision, or ANDEV. We invite you to go to the website and join ANDEV.

OVERSEAS OPPORTUNITIES
DISTANT FIELDS ARE GREENER: SO ARE OUR INVESTORS

Destruction Of Your Assets

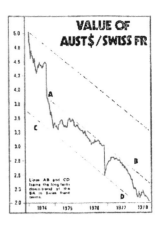

'One Hong Kong Interview' 1977
The author, Bic-Mai (accountant), Carene Klintworth-Tier, Mark Tier

6

THE ALIENATED AUSTRALIANS[8]

Many of these alienated Australians have decided to observe, from a distance, what appears to be the economic destruction of Australia. They all have in common the fact that they know enough about honest economics (not the Keynesian magical myths of 'more government spending' solving all mortal problems). They also share the opinion that Australia will be a great place to return, when some semblance of economic sanity prevails.

Government's only legitimate role is to protect the life, liberty and property of the individual. Just how badly our governments (of both major Socialist Parties i.e., the Labor and Liberals) are fulfilling this role is demonstrated by the development in Australia of a brand-new industry. Perhaps we could call it the 'survival industry'.

This new industry is designed to protect the individual from our 'protectors', the government. Unfortunately, our government's apparent dedication to destroy our money and the rights and independence of individuals has created the very real need for the following services:

- Avoidance of our suffocating levels of income tax.

- Avoidance of over-zealous government re-distribution of resources, away from those who legitimately acquired them by economic means, to those who feel it is easier to acquire it by

---

[8]   Extracts from Interviews with 27 expatriate Australian businessmen, 1977

plunder, i.e., the political method. For the first time in our history the numbers of tax consumers in this country now exceeds those of the tax producers.

- Redistribution of resources toward those of your choice instead of a faceless mass chosen by our political overlords.

- A way of protecting your savings by offering some alternative to the Australian dollar. Charts show that our dollar is already the fourth weakest international currency in the world. There are only three other international currencies being destroyed by their governments at a faster rate, namely, the Italian Lira, the Pound Sterling and the Greek Drachma.

Nobel prize-winning economist F.A. Hayek says "It is bad enough that your government is destroying your money. However, far worse, it has made it illegal for the individual to protect himself from these government policies". In other words, through oppressive Reserve Bank currency controls.

This present situation has created a conflict between our government-made laws and our common morality. Just whose money is it anyway? If it is yours, surely you are entitled to take any possible action to protect it, short of harming any other individual.

This new financial protection service for Australians simply gives us freedom of choice in currency, in much the same manner as we preserve our freedom of choice in any other commodity – other than those commodities where the free market has been perverted by tariffs and other government impediments to free trade.

Australians who travel the world become acutely aware of the difference between the mentalities of the bureaucrats in Asia – it is only money that motivates them – and Australia – where their main motivation is ideological, most of their efforts being directed at tearing down some other man's house rather than building one of their own.

Our Kalgoorlie Chamber of Commerce received a circular from Hong Kong recently, suggesting that Australian businessmen should look at re-establishing (re-locating) or shifting their operations to Hong Kong because "Hong Kong is different to Australia. We welcome businessmen and encourage profits".

This, in a nutshell, explains why we have this new race of Alienated Australians, casting off their bureaucratic manacles and seeking refuge away from our 'concentration camp' commercial environment. Their only sin is that they wish to feel welcome wherever they are and to reach maximum effectiveness in their chosen vocations rather than become submerged in the complex web of devious, self-preservation techniques needed for survival in Australia.

Perhaps when the novelty and temporary excitement of this spy-versus-spy existence in Australia wears off, many more of us will join the Alienated Australians and with sorrow become disinterested observers of the politically co-ordinated, no matter how well-intentioned, destruction of our country.

If we called in a vet to diagnose Australia's ills, on the grounds that it was once a healthy animal which appears to be dying on its feet without any obvious external reason, I'm sure the diagnosis would be quick and the vet would prescribe instant deworming! Even a once-healthy animal can sustain a large and growing population of parasitic worms for only so long, before dying of anaemia.

Rights of the individual

In the U.S.A., the Federal Government tried to close down a farmer's operation because he was selling sheep without government inspection. The farmer replied that he had the inherent and inalienable right to butcher his own livestock and to contract for its sale, with consenting adults, without the benefit or hindrance of a State licence or inspection. The farmer pointed out that the particular Meat Inspection Act, applied only to 'licensed' establishments and to those who have applied for State Inspection Services.

The rights of government

If I live next door to you and you are out mowing your lawn, do I have the right to come over and tell you, that you have to have a license to operate a lawnmower? I don't have that right as an individual.

Now, if I don't have this right as an individual, then I can't delegate this right to government. Government has no inherent rights, only we, the people, do. Government has only delegated rights. Rights which you and I delegate to them.

Freedom

There is not one right, not one freedom, that you have that somebody didn't stand up and either die for, or lose some of their liberty or property, or go to jail for. We have no government on the face of the earth that has continually granted liberty to its subjects. They have always taken liberty away.

That is why Thomas Jefferson wrote in the 'Declaration of Independence' that we have the right to stand up and resist government. I doubt whether there is any secondary school syllabus in Australia which includes the teaching to students of just how the Constitution is supposed to protect us from our government. Nowhere are we taught that government has no right to invade our right to absolute privacy. Are we taught, anywhere, that the coercive agents of government need a search warrant to search for a specific item and that this search warrant covers no other items? Are we taught, anywhere, that the government must always abide by the law?

I can name many examples of two different sets of rules – one for them and one for us.

Am I naïve in believing that the government can best promote law and order by setting a good example? I know enough about the government to know what sort of examples they are setting today.

Our whole present system is based on subservient obedience. It is a system that could not deal with millions of angry, dissident patriots, demanding their constitutional rights.

Imagine if all the employers in Australia simply ceased to continue in their role of unpaid tax gatherers and from a specified date refused to collect the tax from the pay packets of those who worked for them.

The prospect of this kind of thing is enough to give the hardest, meanest, taxman nightmares!

There are many simple and safe ways to start this tax revolt. You have nothing to lose and a tremendous amount to gain by doing this. In fact, you probably won't ever have to pay many of these taxes. Best of all, you would help free yourselves from the ravages of our runaway socialist governments. If you don't participate in the tax revolt then, along with your money, you will be losing your freedom. Freedom is worth what it costs.

As a businessman I object to being cast in the role of an unpaid servant and in fact a thief! If the government wants to rip you off for thousands of dollars in taxes, when you buy a car, I think they should make their own arrangements and directly rip you off instead of doing it in such a way that you don't know exactly how much you have paid in various sales taxes. All you know is that you seem to be paying a lot more money for a car that costs much less in other countries. [When I purchased a Lexus for $130,000 in 2005, I paid a total of $30,000 in taxes.]

I also object to being forced to rob by the laws of this country, laws that are supported to protect people. The law states that if you are working for me, I must stick my fingers into your pay-packet and take out between 30 and 60 per cent of your pay. This, I must put together with other money that I have to steal from other pay-packets, complete the necessary documentation and send the payments regularly to the government. I think that the government should be honest enough to come to you, on a face-to-face basis, instead of involving me in this whole sordid affair.

No being content with all this subterfuge and nonsense, they fine me, as the employer, for creating employment. Every month they fine me what they call payroll tax. The fine is imposed only because I employ people. The more people I employ, the more tax I pay.

In our society the prizes appear to be going to the wrong people. If I wish to get the prize of not being fined, then all I have to do is go out of business and cease employing people.

Now we have got a problem in Australia called unemployment and it is only going to get worse as many employers realise the benefits of shrinking their enterprises. Sometimes this is the only way to survive. Accompanying this continual business shrinkage will be a decline of the stock market because of the continued trend toward 'redistribution' away from profits towards wages.

The politicians are naturally concerned but, as usual, rather than taking any steps to restore incentives, they can only think of solving the problem by more controls and centralised edicts. The real solutions to the problems are to be found only through fewer controls and more freedom of choice for each and every individual in Australia. The individual can only be left free, to choose the level of their efforts, if their rewards fluctuate with the value of the services they personally provide. That value being determined in the marketplace and not by politicians and bureaucrats.

If individuals' incomes are increasingly determined by politicians and bureaucrats, they lose not only incentive, but also the possibility of finding out what they should do in the general interest. When they cannot know, themselves, what they must do to make their services valuable, then they must be commanded. This is when our politicians and bureaucrats leap to the challenge of justifying their existence. At that point people walk away from investment.

On a more positive side, Australia is providing the world with the best training ground for management. If you run a business efficiently, plus overcome all the bureaucratic hurdles in Australia, then you will be an outstanding performer in other countries. Overseas companies enthusiastically recruit Australians. There is a strong export of executives from Australia. They strike a very sympathetic cord when they attract Australians with phrases such as, "by working in our country you will find essential differences" and "unlike Australia, we welcome businessmen and encourage profits".

Quotations from my interviews with 27 expatriate Australian businessmen –1977

[Note: I have severely edited the 27 taped interviews for the following reasons:

- Many of these individuals have since returned to Australia and as at 2019 their comments may not be regarded as helpful. One example being the late Mr W.R.A. (Bill) Wylie who, after returning to Australia in 1987, became one of Western Australia's respected business leaders.

- I am no longer in personal contact, so am unable to seek permission.

- In the interest of brevity.]

Conversation with John, an Australian engineer, now based in Hong Kong.

RBM: John, what brought you from Sydney to Hong Kong?

I think, to me, having lived through the period of post-war England where there was complete reluctance, on the part of anybody, to do a day's work and the featherbedding by governments, it seemed to me to be occurring again in Australia and, I regretfully say, largely as a result of the shop stewards and the pommy migrants who came out to Australia. I had the opportunity of coming out here (to Hong Kong) with another company, an Australian company, which had a contract in Hong Kong and within a matter of days I suddenly felt the enthusiasm of people who are prepared to work and who are living free – 4¼ million Chinese people, and although they are living under very adverse conditions, they seem to have a more hopeful attitude to life than the people living in Sydney who have far greater comforts but less *joie de vivre*, if I may use that expression.

Probably, to sum up my attitude, I liked living in Australia, I liked the spaciousness of it, I would quite like to go back if conditions change there, but in the meantime for anybody who has been engaged in private enterprise, in its proper sense, Hong Kong is the tremendous challenge, nobody gives you anything. They have to go out and get it themselves. But if you do make it then the rewards are there – in money, low taxation and in the satisfaction of achievement.

Conversation with Philip, an Australian lawyer in Hong Kong

RBM: Philip, what really triggered your move out of Australia to Hong Kong and when did this happen?

Well, a job opportunity was put to me. Compared to the opportunities as I saw them at the time in Australia, I really couldn't refuse it, from a personal advantage point-of-view and also from the monetary point-of-view. It was becoming increasingly difficult with tax burdens, based on the Australian tax scale. It was becoming increasingly difficult, not just to make ends meet, let alone try and get ahead and build up some basis of personal financial security. I cannot see, at the moment, how a young couple in Australia, for example, is going to, within the next ten years, have the opportunity to achieve what has world-widely been known as the great Australian dream and that is owning their own home. They are burdened with very high tax rates, even though both of them may be earning an income. It is becoming more and more difficult for them and this is one of the reasons why I looked at my position. I saw the opportunity of a chance for five years to work in another environment, an environment where people work and are paid in relation to their work and their ability and not, as we are finding in Australia today, where people are paid and if you are lucky you get some work out of them.

RBM: Philip, in Australia, the Australian government and their dreaded tax enforcers always refers to Hong Kong as a 'tax haven' and in a fairly derogative sense. How do you feel about this term being used in relation to Hong Kong?

Well I think it is quite ludicrous really because people in Hong Kong do pay taxes. They pay taxes at what I consider to be realistic rates. The fact is that personal income tax cannot exceed a maximum of 15 per cent and the corporate tax, at the moment, is 17 per cent on income earned in, or derived from, activities in Hong Kong and offshore income is non-taxable. This, I think, is a very realistic way of approaching the tax requirements. It still enables people to have an incentive to work harder to make more money and this obviously reflects in the productivity that is achieved in a place like Hong Kong where people work industriously, and the rates of productivity continues to increase.

Hong Kong is very dependent on imports of raw materials. It is dependent on its import of basically everything and we are still able to achieve what I consider to be an excellent annual growth rate between six and seven per cent, only achieved because of the fact that people are working, and productivity is increasing. They are doing this because the incentives are there and they know the extra dollars that they earn are still only going to be taxed at around 15 per cent.

RBM: I have met a great number of Australians during this trip up here and none of them have anything adverse to say against Australia itself, as a country. In fact, they are still excited about Australia's potential and saddened by the way the country is just being ground to a halt. Any derogatory comments they have to make, of course, are only toward our political and bureaucratic 'overlords'. I am wondering, with all these ex-Australians gathered here in this one centre, is there anything that could be done by you all here to assist Australia from something of a distance?

Well there are many things, I suppose, that we could do. It is a matter of whether anybody is prepared to listen to them when we do them and that is part of the problem Australia faces today. I think that, as Australians, working here, we know what can be done and what can be achieved. There are sufficient of us here to do something like this. We could be working together, as a unified group, to try and make governments and others in Australia aware of the problems that some of these decisions are causing us in our business activities. I mentioned how difficult it is, in some places in Asia, to sell Australian products because of the tariff barriers that have been put up against these countries going into Australia. We need to take a more active part in lobbying and trying to make governments see reason for the actions they are taking and reverse them. It is not easy, the lobbies in Australia, at the moment, tend to be very strong as to protectionism and they seem to be holding sway with the government. Now whether as individuals we can do this or whether we need to form into a group to do it – well, that is something which we have to decide. But quite frankly something does have to be done and something has to be done in the near future.

RBM: In your almost three years of living in Hong Kong, how

many times have you had the opportunity to revisit Australia and what has been your feeling upon returning to Australia and again on your feelings when returning to Hong Kong?

Well, I have had the opportunity to go back on quite a few occasions. I think the first occasion gave me the greatest jolt. It was after some six or seven months here and I had become used to the conditions of business in a place like Hong Kong, as well as other parts of Asia, where people work hard and one can get things done. If you are energetic and you like to have a reaction with a similar vein of people you are dealing with, well this is the place you can get it. It appalled me, quite honestly, when I was first back, at the rapid change in peoples' attitude towards work. I suppose this had been going on around me for quite some time, but I hadn't recognised it, while living there, before I came up here. However, having been away for six or seven months and going back it became far more noticeable that it was difficult to get people to do things. People tended to be 'nine-to-fivers'. If you wanted something done, okay, it will get done in sometime and it will be 'our time' not 'yours', not when you want it and so on. While here, in Hong Kong, we find if we want to get something done it can be done within a matter of hours. If not within hours, then within a few days and particularly if you are dealing with people in business. You have quick decisions made. This is important in Hong Kong because of the competition. Again, if the person you are dealing with doesn't make it quickly, then one of his competitors will beat him to it. I must confess I left Australia, after my first trip back there, feeling very depressed and I was delighted to come back to Hong Kong. I arrived back, on a Sunday, knowing that on the Monday I was getting back into an environment where I was going to be happy working. I knew I could achieve something and achieve it quickly and effectively, without having to cut through the attitudes of work, the red-tape and so on, which Australia seems to be building up at the moment. In fact, I was so pleased to return that as I stepped off the plane, I felt like kissing the ground.

I would say, however, that I don't wish to be considered to be one of those good Australian 'knockers'. I have had the opportunity, over the past 15 years, of seeing many countries around the world. I still consider Australia to be one of the greatest countries in the world and we do have so much going for us. I certainly intend to return there, when attitudes change a little and when I feel that I can go back and get satisfaction out of working in an environment similar to Hong Kong. What does disturb me, greatly, is sitting up here, as an Australian, seeing people in this region looking at Australia from the outside and laughing at us saying, 'look you have got it but you are a great bunch of fools because you don't know how to handle it and what to do with it' and this is very disturbing. If you look at it from that point-of-view, you might ask yourself how long it is going to be before Australia's fate will be decided from outside Australia.

RBM: Many Australians are getting to the stage where they fear picking up the morning or evening papers because our newspapers read like a catalogue of disasters, but I don't get this feeling here in Hong Kong. I am quite enthused, after having put a newspaper down, about just what is happening in terms of actual world coverage. I wonder if I could just ask a quick comment on what the essential difference is in our media coverage?

Well, I suppose you could say that a lot of Australian press is based on the model of 'sensationalism sells newspapers'. However, that is not the fact here really. They don't come anywhere near the same style as the sensationalist newspapers you have in Australia. Therefore, we don't have the so-called disasters blown up out of all proportion. I think it is probably fair to say, too, certainly we don't have anywhere near the same problems in Hong Kong that you have in Australia at the moment, for the papers to latch on to. Generally speaking, the standard of the media in Hong Kong is exceptional. It is servicing diversified requirements here, people from all over the world. It is keeping them in touch with happenings all over the world. It has a very extroverted approach towards reporting. Whereas, in Australia, I feel, in keeping with a lot of other attitudes, the press is becoming terribly introverted in what it is doing and what it is saying. In Australia we are far more interested in what is happening in our little neck of the woods. People need to get out and look at the real world because with the attitude we have in Australia at the moment people are just quite unaware of what is happening in the real world. Whereas, in Hong Kong, the real world is all around us. The world of commerce, the world of politics and what happens on the other side of the world, can very dramatically effect what is going to happen in Hong Kong. Perhaps not just politically but certainly economically and therefore, the press keeps catering for this need, as the Australian press should be catering for this in Australia but unfortunately at the moment it is not.

Neil, an accountant in Hong Kong.

RBM: Neil, you have been in Hong Kong now, I'm told, for four years and you mentioned you re-visited Australia every three or four months. This gives you a pretty good yardstick on the state of affairs back there compared to Hong Kong. What about some comments on just Australia, as you see it, from time to time on your visits back?

Over the last three or four years I have been in Hong Kong, there has been a noticeable change in the business environment in Australia. I think this originally started when the Labor Government first got into power. Since then I have found there have been a lot of Australian companies looking overseas to other areas where they might be able to diversity their activities. This is due to the, at home, labour problems, the size of the market, problems of distribution, distance between markets and this sort of thing. I don't really notice a great difference in the country each time I go back. Mainly because I'm going back every three or four months and it is not really sufficient time, on each visit, to notice a great difference. I do think that the country, generally, is not progressing at the rate that it might otherwise progress for a number of reasons. One would be the 'Lucky Country' philosophy which still seems to exist in peoples' minds. This is generalising a bit but is supported by the resources/mining sector and the wool/wheat type of industries. Manufacturing industries generally can't be competitive with the rest of the world. Some of the reasons are high labour costs, lack of market size and problems with distribution which I mentioned before. Another reason for the changes that are occurring or the lack of incentive in Australia would seem to be the government not really giving sufficient leadership to the country, and taxation problems with taxation being relatively higher than a lot of other countries. People are spending a lot more time trying to avoid tax rather than paying their tax and getting on with the job of producing.

Mark Tier, an Australian economist and publisher of World Money Analyst.

RBM: Mark, just when did you leave Australia and what caused your departure?

I left in August 1976 to do a PhD at UCLA in Los Angeles but that was only part of the reason for leaving. We always leave to do something. If you leave a country, you leave for a reason. You are going somewhere to do something and I left because Australia was getting me down. What has happened to me since I left? I'd moved my newsletter business to Hong Kong but was living in LA. Something had to give. So, I quit the PhD after two months and came here to Hong Kong where I now work. As one of my fellow students put it, like her I'd been in the real world for too long. She didn't last much longer than me.

RBM: Mark, you were writing your investment newsletter before you left Australia. Was that your main activity in Australia?

Yes, and writing articles for various magazines. I was involved in the Workers Party and I still write the *World Money Analyst* for

Australians. We have a special Australian edition. So, I am still in touch with Australia.

RBM: Were there certain aspects of Australia that tended to depress you, because I have gained the impression this has been the case for many other Australians I have spoken with in Hong Kong? What was your main dissatisfaction with the situation in Australia?

Well Australia is just 'going down the drain'. There are two things, firstly, Australia is cut off from the rest of the world, never really knowing what is going on anywhere else. You don't really feel connected or plugged in. Secondly, opportunities are so limited and recognition of anyone with talent is limited. You have to fight, it's an uphill battle. The Australians make peanuts when they could go to the United States and work less hard and make triple or ten times the amount of money.

RBM: Mark, tell me what really causes Australia to be remote and cut off and, as you say, "not plugged into the rest of the world"? Is it related to our media or attitudes or what?

Not so much the media as the 'Australian' media which, in my mind, is one of the better media around the world. The American media is the worst but nothing compared with the London newspapers. So, it's really a lot to do with economics. There hasn't been much work done on this aspect and the effects of tariffs and other barriers to trade. However, there are two things which come to mind. The first is the effects of tariffs which isolate Australia from the world market and cause increases in the cost of the raw materials. The result of that is that the new products and new ideas that are happening in the U.S.A., Europe, Hong Kong or wherever, might not get to Australia for 3 or 10 years later, if then. It is just the little things that can illustrate how the effects are being cut off by tariffs. There are products which had completely disappeared from the Australian market and are now just coming back on, but at ten times the real price and as luxury items!

Another example is the little Sony tape recorder. This particular model is just not in the Australian range because they say, in Australia, they would have to retail at $300 which is just too expensive for the Australian market, so this is another example of their restrictions of freedom of choice, as a result of trade barriers.

The second point is with travel. Travel in and out of Australia is so expensive. It is out of the question for most people to go somewhere for a week or so, whereas if you are living in New York you can go to London or Paris for a week – anywhere in Europe. The only place here you can go is somewhere overseas. Of course, there is nowhere to go in Hong Kong. It is so small and there are so many people. If I rented a car, for a weekend, I could go everywhere in Hong Kong twice. However, Australia should be a great country but it is wasting away and I don't see anything happening, at present, to change the way it is going economically. Prime Minister Malcolm Fraser came in on a platform of free enterprise and he has done nothing. He has put half of Whitlam's schemes into effect and done nothing to change the fundamentals whatsoever. I see, right before my eyes, Australia going a lot the same way as England. Australians don't know it because they are cut off from what is happening elsewhere. They can't see what is happening in the rest of the world.

RBM: Mark, what is the main function of your monthly newsletter – *World Money Analyst*?

Its main function is to help people survive.

RBM: Survive against what?

Against government. To help people wherever they may be to get their assets out of the hands of governments and into their own hands to avoid taxes and government regulations.

RBM: So, you see government more as a predator rather than a protector?

> Yes, that is true about every government in the world and there are a few places here and there that have no government to speak of and Hong Kong has got that sort of government.

RBM: The government is almost invisible here?

> Well it is to the passer-by because the contrasts are just so enormous. When you have been here for a while you realise that although the government is incredibly smaller than Australia it is trending in the same direction. What's amazing about Hong Kong compared to Australia is that there's no politics here! It's such a relief.

RBM: Mark, how does your shift of base of operation to Hong Kong effect you in the writing and publishing of your newsletter?

> For a start I am writing for the world and not just Australia alone. We do two editions. One a special Australian edition with Australian news, comments and analysis. Then there is the international edition that goes to the rest of the world. When the circulation builds up, in the future, we will probably go into a lot more geographical editions, like *Time Magazine*, with 66. However, I think my perspective has changed quite a lot. I don't feel closeted into the narrow straight perspective. I feel less emotionally involved with what is going on in Australia. What I have noticed, with a lot of people, is that they get terribly uptight about what the government is doing. All they get is ulcers because no individual can have the power to change the world or change a country. The best course of action is to start protecting yourself and putting yourself into the position where you are immune from the government and then, when you are immune, like you haven't got the disease anymore, you are in a much healthier position to start fighting. That is one of the things I do in the newsletter. I am hardly a friend of Prime Ministers Fraser or Whitlam or

anyone else in Canberra. Of course, the other thing is whatever I was selling, in Australia, in terms of circulation I sell 20 times that amount in the U.S.A. plus another 20 times that in the rest of the world. So, in terms of profitability, the potential is many thousand per cent increased.

RBM: I notice the postal costs are about 25 per cent of what they are in Australia and other parts of the world, so that must contribute to the efficiency of your operations here in Hong Kong. What other similar efficiencies do you find here?

Well, in fact, we are spending about twice as much on postage because of the weight & sending via airmail. It takes about the same time to get there. It is difficult to compare the newsletter with what it is now because it is a much bigger production but certainly on the cost side, every cost is lower, paper, printing, labour. Everything, except rents.

RBM: No sales tax on the printing of paper either?

Correct, no sales tax on the printing and, of course, no income tax because off-shore income is not taxed, only income locally derived.

RBM: I guess you avoid selling any newsletters locally.

Well it is only 17½ per cent anyway. Although, here again, the perspectives are different. 15 per cent maximum income tax sounds pretty good but to my secretary it is dreadful because she doesn't come into the 66 per cent of the population who pay no tax at all.

RBM: You appear to be having a lot of fun providing this medical service for those who are forced to live in a sick economy. I hope the medicine is accepted and solves the problems that we are confronted with in Australia. Mark, you went back to Australia in January. That would have been four months after you had left. What were your reactions? Australians usually have a very definite reaction when they lob back in their country after a few months away. Did you have any first impressions that you would like to relate?

Yes, it was really great to be back because I am an Australian and always will be. Oppressive. Four months isn't really a long time and in that time nothing had changed. The issues were the same, the problems were the same and the solutions were nowhere in sight, so far as coming to the political arena. What was even more depressing was one day, a couple of months back, we were a little short on copy for the Australian section for the newsletter. I went down to the shop here and bought a couple of *Financial Reviews* and discovered nothing had changed. That is one of the reasons, in a sense, why I was glad to leave Australia. I was starting to get bored and monotonous, like writing the same thing every two weeks but just changing the dates.

There are very few times I feel ashamed to be an Australian. It's not that I felt particularly proud about it either. Mostly, I simply figure that it is a fact of my existence. However, a few days ago a friend of mine sent me a few $50 notes as payment of a debt, so I went downstairs to the bank to exchange them for Hong Kong dollars. 'We only take US dollars', they said. So, I went next door to another bank. They looked at these notes as though there were something strange about them and directed me to a corner money-changer down the road. I really felt as though I was carrying around some exotic currency from some communist nation which no-one had ever heard of before. And were not interested in hearing about now. I offer this little experience to you for what it is worth.

Barry, the Hong Kong Packaging Manager for Coca-Cola.

Since moving from Australia to Hong Kong I am now banking (saving) a total of my previous Australian salary. That's after leading a good life with my wife and family. All due to the favourable tax scale.

It was easy to adjust to Hong Kong life. As a matter of fact, when I returned to Australia, in March, for four days, I received such a jolt as all my friends were up to their 'asses in alligators', the way I had been before I left. My feeling when I returned to Hong Kong was that I was coming home!

Bob, an engineer and MBA from Australia.

I don't even go back to Australia any more. Two main problems that drove me out of Australia were:

- Most manufacturing industries are just not viable on sound economic grounds and need propping up at great expense.

- The largest government, in the world, in terms of proportion of population. I can refer you to a recent article in the *Asian Wall Street Journal*.

Luke, an Australian lawyer, now working in Hong Kong.

Looking back at Australia, I've become aware that the only way to survive today is to take advantage of the tremendous confusion about what is legal and what isn't, to maintain a low profile, crawling on our bellies to escape official notice. Cautious, ponderous timidity is the style of doing business in today's Australia.

An Australian surveyor, now working with a Hong Kong property developer.

Physically Australia could still be called a 'Lucky Country' capable of much. However, the disincentives of socialism (rapacious tax rates) have made it virtually impossible to legally keep what one earns.

Bruce, an Australian architect, in Hong Kong.

I feel considerable apprehension at having to readjust to life in Australia where for one to be his own man is synonymous with being a rebel. The refreshing thing about countries where some degree of personal and economic freedom exists is the total compatibility of being 'your own man' with nothing to rebel against and channelling physical and mental effort, instead, into creativity and productivity. In Hong Kong, for instance, it is possible to experience the complete spectrum of human sensitivities. This is an impossibility in Australia as we appear to be submerged under an avalanche of apathy.

Here are several more samples of comments made during my interviews.

Hong Kong's assumption is "that in a free-wheeling, healthy economy, the number of people requiring legitimate welfare assistance will be minimised to the stage where voluntary (non-government) charity can handle the problem." That assumption appears to be well based.

- "The Hong Kong Government does not see their role as picking you up if you stumble but to remove every possible obstacle so that stumbling is minimised."

  (That attitude reminded me of the Leonard E Read (FEE) quote: "The politician's idea of helping the poor to become helpless is no act of kindness.")

- "In Hong Kong the prizes appear to be going to the right people, unlike Australia, where it would appear that a social worker rates higher in status than a geologist or engineer."

- "Australian hand-outs may be more liberal than Hong Kong's, but this is no real advantage to those who do not wish to live at another's expense, for those who wish to assume responsibility for their own lives."

- "Hong Kong is a place where one can be happily married and yet not let one's marriage interfere with one's private life."

Australian Taxation Office 'kangaroo court' workings revealed

ROBERT GOTTLIEBSEN
Follow @RGottliebsen

By **ROBERT GOTTLIEBSEN**, BUSINESS COLUMNIST
8:03AM JUNE 20, 2019 • 55 COMMENTS

THE AUSTRALIAN

ATO GST powers should be a worry for business

ROBERT GOTTLIEBSEN
Follow @BGottliebsen

By **ROBERT GOTTLIEBSEN**, BUSINESS COLUMNIST
7:53AM JULY 22, 2019 • 31 COMMENTS

BUSINESS   SMALL BUSINESS   FINANCE

Inquiry finds 'excessive' debt recovery action by ATO against small business

By Cara Waters
April 28, 2019 — 9.30pm

Recent articles on the behaviour of the Australian Taxation Office reminds me of my 1970s experiences.

7

ADVENTURES IN TAXATION

(Help starve a feeding bureaucrat)

1978

"What is very easy? To advise another.

What is very hard? To know yourself."

— Thales, the first philosopher.

Arthur Seldon on Taxation – how times have changed

When sorting through some old files, I came across some notes from a visit to London in 1984 where I sat down with Arthur Seldon. Arthur was one of the two intellectual geniuses, the other being Ralph (Lord) Harris, selected by Antony Fisher to operate the Institute for Economic Affairs, established in 1955. Arthur was, on the day, as usual, surrounded by notes and ear-marked books as a fellow-attendee at the second Libertarian International Conference, 1984, in London. I only had to ask one question to get him started. I referred to the oppressive level of taxation that was burdening the U.K. (and Australia) at that time.

My question:

Historically the tax collector used to collect the taxes and they could be clearly identified as the 'villain'.

In more recent times weak business leaders have allowed themselves to be manoeuvred into becoming unpaid tax collectors making it less clear just who is taxing us.

Ayn Rand was probably referring to weak business leaders when she said, 'It cannot be done to you without your consent, if you permit it to be done, then you deserve it.'

Arthur, the average taxpayer no longer has the knowledge of how much tax they are paying or how to go about questioning this level of tax so my question to you is:

Have we left it too late to revolt?

Arthur's comprehensive reply (without even pausing for breath) was as follows:

So, if I say, that time is on our side, I also have to add that we must give it a bit of a tug and I think we can do it in a number of ways. The first is in our tax structure.

I'm in favour of a withholding tax. I'm in favour of taxpayers withholding their taxes until they are satisfied with the services that Government has supplied them with and although I can offer you no advice, which is outside the law, I can at least paraphrase a book you should buy called *The Bureaucrats and How to Annoy Them*. The advice that this man gives is very perceptive and he has to hide behind a pseudonym[9]. He is a noted scientist who broadcasts quite a bit. Some of the things he says are as follows:

1. When writing to your tax gatherer, be vague; write in illegible longhand and at wearisome length.

2. Use plenty of incomprehensible jargon (after some days send another copy to confuse him).

---

[9] It's pretty clear that the author was Patrick Moore, the acclaimed amateur astronomer and writer.

3. Write back to his non-existent colleague.

4. Give incorrect reference numbers.

5. Request a reply to a letter you have not sent.

6. When you do pay, send the wrong amount.

7. Staple your cheque in the centre to jam their computer.

8. Better still, crumple your tax demand, iron it out and refold it in a different place (guaranteed to defeat their electronic receipting).

9. Use wrong dates (but keep a note of the right ones).

10. Stick the stamp in the top left-hand corner of the envelope, or better still, in the centre.

11. Query **all** your tax assessments.

12. Send the tax gatherer your detailed form, demanding information about his authority to levy these taxes.

13. Rub candle grease on the space marked 'For Official Use Only'. If you are short of candle grease, use hair oil.

14. Observe in general an attitude to your taxmen (your tormentors) of divine neglect.

My final reason for thinking that time is on our side is that the **power of ideas** in the end will dominate.

Keynes was wrong in so many other ways that I hardly like to add one more. He was wrong when he said simply that **ideas** would dominate.

So was Marx wrong when he said that **interests** were the ultimate force.

But John Stuart Mill was right when he said: "Ideas would determine action, provided circumstances conspired to make them timely", and I think that the time is almost right. Time is almost right because the ideas that we have discussed here are being married to a period of technological change which will undermine the power of government and also to a climate of morality that will see that it is proper for men to take a long-term view of their future and not the short-term animal myopia that our socialist enemies now teach.

Today, 35 years later, I look back at Arthur Seldon's optimism and observe that taxpayers did not seize the technological opportunities to which he referred. So, I conclude with my verse.

Nothing much will change it seems,

till we get our knees up off the floor.

Stop begging for protection from competition –

That's just a perversion of the law.

The problem's solved if we all stand up

and **decline** their invitation.

Let them collect the taxes

If **they** want to run the nation!

TAX — VALUE FOR MONEY?[10]

Well, fancy giving money to the government.

Might as well put it down the drain.

Fancy giving money to the government.

Nobody will ever see the stuff again.

Well, they've no idea what money's for -

Ten to one they'll start another war.

I've heard of a lot of silly things, but Lor'!

Fancy giving money to the government!

– Sir Alan Patrick Herbert

It all started when I was asked to lead a discussion on Taxation.

The title of the discussion, "Tax – Value for Money", wasn't chosen by me. When I was first asked to speak on it, it sounded like a bad joke. How could you possibly get any value out of being robbed by the tax arm of the Australian Government and then having this money used against you?

In saying this, I am not suggesting that all politicians and bureaucrats are crooks and misappropriating your money. Not all of them are crooks and many of them are well-intentioned, it is just that the actual results of most things they do with your money is the exact opposite of their hopes or intent.

That's why I personally feel that every dollar you let slip through into their hands is doubly disastrous.

First, it reduces your economic freedom by that dollar amount, and renders you a 'Tax Slave'. When Federal income tax was introduced to Australia in 1915 the average income earner paid less than 5 per cent tax. That means that Australians at that time had 95 per cent economic freedom of choice on how they spent their own earnings.

---

[10] *Warning! Written in 1978, so the following comments should, in no way, be taken as advice. It is reproduced here only for entertainment value and an historic record of the era where it was possible to speak freely without fear of reprisals.*

Today, with personal income tax running at upwards of 33 per cent, you have 67 per cent of your economic freedom left and if you take into account all the other indirect taxes, tariffs, etc., the average Australian is probably left with only something like 30 per cent of his economic freedom.

When you look at this ratio of 70 per cent slavery and 30 per cent freedom and reflect that it represents a continuing trend, it probably explains the general concern of many Australians today.

Second, for every dollar you put into the hands of the enemy (so to speak) you have increased his power to act against your interests by launching into more of these well-intentioned, but usually disastrous, programs. (e.g., the Australian Industry Development Corporation, Department of the Media, etc.)

I have ample evidence for my strong views about the difference between the results and the intentions of governments' actions and my bulging files on this topic are available for any researcher.

The nature and kinds of 'market failure' have been analysed by economists since the time of Adam Smith 200 years ago. The nature and kinds of 'government failure' have been discussed only in the last 15 or 20 years. That may be why some economists, concerned about the imperfections of markets (in maximising output and distributing it equitably), have been inclined to argue that it is necessary only to demonstrate that markets are imperfect for them to assert their case for replacing them by government intervention.

Fortunately, many other economists can put forward much more rational proof that freedom doesn't have to be perfect to be the best choice around. Most of us have become a lot wiser over the past few years, in the sense that we realise that far too often the government becomes part of the problem rather than part of the solution.

Cost savings, like charity, should start at home and if governments sincerely wish to balance the budget and bring to an end government-created inflation, then that government must take positive steps itself, instead of talking about other people taking positive steps.

The thickest file in my filing system is titled 'Contradictions – Between Government Intent and Results'. Incidentally, one of my thinnest files is headed 'Legitimate Roles for Government Involvement'.

Let's look at a few clarifications and definitions:

'Legitimate Role for Government'. Since government is organised force, it can only consume or destroy. As it can rarely produce or create, it should be limited to:

- Keeping the peace.
- Restraining all actions destructive of human creativity.
- Invoking a common justice.

That is all!

Three major systems of government

Socialism: The State owns you and the means of production. Here a rigidly planned economy causes gluts and shortages of goods and labour with no personal or economic freedom (the former is the corollary of the latter).

Fascism: You own the means of production, but the State, by regulation, price control, wage control and punitive taxation, enslaves you and reduces investment. This controlled economy again leads to gluts and shortages of goods and labour and inevitably to the loss of economic and personal freedom.

Libertarianism: You own the means of production and yourself. The state exists to provide you protection by the police and defence forces so that one man may not, by force or fraud, coerce another man. A free market economy, with the 'consumer the king', free of government interference, follows the law of supply and demand in goods and labour with gluts and shortages ironed out "as though by an invisible hand" (Adam Smith). Personal wealth and freedom are maximised. Democracy today has adopted the socialist and fascist concepts, with ultimate economic failure and varying degrees of loss of personal freedom, all over the world. The outstanding exceptions have been the U.S.A. under Thomas Jefferson and more recently West Germany under Ludwig Erhardt. The U.S.A. and West Germany are outstanding economic success stories which promote personal freedom. However, there are signs that a decline is beginning as socialist policies are increasingly introduced.

Free-Enterprise

The freedom of an individual to set up enterprises. The freedom of an individual to engage in any activity so long as he uses only voluntary methods of getting other individuals to co-operate with him. Free enterprise = willing exchange. (If you're curious to see how far we have moved from the basic concept of free enterprise, just consider how free anyone is to set up an enterprise – whether it be a bank or a taxi-cab. You need a government licence for almost everything, including voluntary transactions on Sundays.)

Tax: A payment made to government for servitude rendered.

Tax Resister: One who keeps his own money. One who cheats a blood-sucker. Someone who is prepared to help starve a feeding bureaucrat. One who follows the Christian ethic and renders unto Caesar only that which is Caesar's (pay for what you use and regard any surplus contribution as a voluntary gesture).

Tax Payer: One who renders unto 'seizer', that which is not the 'seizer's'. One who feeds the mouth that bites him. A beast of burden, specifically, an ass!

Poorhouse: A home for conscientious taxpayers. The ultimate tax shelter.

Reflecting back on the title of this talk, after worrying about it for a while, I realised that there is a way to get better value for our tax dollar, it seems also that it may be the ONLY way.

To appreciate this, let us first examine our current position.

> Every government intervention into peaceful private activity tends to make things worse rather than better.
> – Ludwig von Mises

> The more laws and restrictions there are,
> The poorer people become.
> The sharper men's weapons,
> The more trouble in the land.
> The more ingenious and clever men are,
> The more strange things happen.
> The more rules and regulations,
> The more thieves and robbers.
> Therefore the sage says:
> I take no action and people are reformed.
> I enjoy peace and people become honest.
> I do nothing and people become rich.
> I have no desires and people return to the good and simple life.
> – Lao Tsu, 6th century B.C.

> Don't swallow all the rubbish talked about the Company being a separate entity being adequate reason for tax-strangling the enterprise and snuffing out vital initiative.
> – Eric Risstrom, Taxpayer's Association

Don't be perturbed by the thought of others evading tax. If they paid their tax in full, you wouldn't pay any less. The government doesn't operate along sound economic lines like this. They spend it even when they don't collect it. They just print the difference and that's why we have inflation.

Is there a problem?

The *Encyclopaedia Britannica* defines taxation as "that part of the revenues of a State which is obtained by the COMPULSORY dues and charges upon its subject". That is about as concise and accurate as a definition can be. In that definition, the word "compulsory" looms large, simply because of its ethical content. The quick reaction is to question the right of the State to this use of power.

On this question of morality there are two positions and never the twain shall meet:

A. Those who insist that political institutions stem from 'The Nature of Man' thus enjoying a form of divinity. These people can find no quarrel with the concept of compulsory seizure of the property of individuals.

B. On the other hand, there are those who recognise the primacy of the individual, whose very existence is his claim to inalienable rights. Their position is that the compulsory collection of taxes by the State is merely the exercising of power, without regard to morals.

The latter is my position: I believe in the rights of individuals and feel that the State's compulsory claim on my life is morally dubious. Rather than being biased, I feel that I have looked at both sides and have taken the trouble to understand my basic premise on this subject.

If we assume that the individual has an indisputable right to life, we must concede that he has a similar right to the enjoyment of the products of his labour. This we call a property right, which is a *basic human right*. The absolute right to property follows from the original right to life because one without the other is meaningless; the means to life must be identified with life itself. If the State has a prior right to the products of one's labour, his right to existence is in doubt.

We object to the taking of our property by organised society, just as we do when another individual simply commits the act. In the latter case we unhesitatingly call the act robbery. It is not the law which in the first instance defines robbery, it is an ethical principle, and this the law may violate but not supersede.

If, through sheer necessity of living we have compromised and agreed to an existence under some set of arbitrary rules and, through long custom, lost sight of the immorality involved, we should remember that the principle has not been obliterated. Robbery is robbery, and no twisting of words can make it anything else.

Any historical study of taxation points inevitably to loot, tribute, and ransom – the economic purposes of conquest. This has been so right from the early gangs who 'protected', for a forced fee, the caravans going to market, and later included the Danes and the conquering Romans. The only thing that time has changed is that constitutionalism has somewhat diffused political power and made it slightly less obvious that these taxes served only to keep a privileged class in comfort and to finance schemes to retain them in power.

Again, when you get right down to it, taking money from one person and simply giving it to someone else is theft, morally if not legally. Laundering the money by having the government act as agent for the transfer doesn't improve the situation ethically, even though it makes it apparently legal.

I haven't seen anyone explain this concept of legalised plunder better than Frédéric Bastiat, who almost 150 years ago stated:

> But how is this legal plunder to be identified? Quite simply …
>
> See if the law takes from some persons what belongs to them and gives it to other persons to whom it does not belong. See if the law benefits one citizen at the expense of another by doing what the citizen himself cannot do without committing a crime.
>
> Then abolish this law without delay, for it is not only an evil itself, but also it is a fertile source for further evils because it invites reprisals. If such a law – which may be an isolated case – is not abolished immediately, it will spread, multiply and develop into a system.
>
> The person who profits from this law will complain bitterly, defending his 'acquired rights'. He will claim that the State is obliged to protect and encourage his particular industry; that this procedure enriches the State because the protected industry is thus able to spend more and so pay higher wages to the poor working man.
>
> Do not listen to this sophistry by vested interests. The acceptance of these arguments will build legal plunder into a whole system. In fact, this has already occurred. The present-day delusion is an attempt to enrich everyone at the expense of everyone else; to make plunder universal under the pretence of organising it.
>
> … Socialists like all other monopolists, desire to make the law their own weapon. And when once the law is on the side of socialism, how can it be used against socialism…
>
> The choices before us.
>
> This question of legal plunder must be settled once and for all, and there are only three ways to settle it:
> - The few plunder the many.
> - Everybody plunders everybody.
> - Nobody plunders anybody.
>
> We must make our choice among limited plunder, universal plunder, and no plunder. The law can follow only one of these three:

- Limited legal plunder.
- Universal legal plunder.
- No legal plunder...

Under our existing system of 'universal legal plunder' the collection of taxation falls into two categories, direct and indirect.

Indirect Taxes

Indirect taxes are so called because they reach the State by way of private collectors (the term derived from the days when direct taxes were paid directly by the taxpayer and not collected, as today, by private or company collectors in the form of the Pay As You Earn (PAYE) system.

Indirect taxes are attached to goods and services before they reach the consumer and it will be seen that they are in fact a 'permission-to-live' tax. There are few items in the market place to which these taxes are not attached, hidden in the price and, under compulsion, you either pay or go without.

It is the very inevitability of this tax that makes it so popular with most governments. Many of us are astounded by the disparity between the cost of production and the charge for 'permission-to-buy'. To explain this, someone has put the number of taxes accumulated in a loaf of bread at over 100.

One of these indirect taxes, sales tax, requires the unpaid labour of the business community in collecting and forwarding the taxes to the government.

The high cost of tax collecting for the government has another unfortunate effect (intended or unintended?) These costs, readily undertaken by large businesses, impose a disproportionately heavy and often crippling cost on the small employer. The large business can then cheerfully shoulder the cost knowing that his small competitor bears far more of the burden.

Payroll Tax is another tax I regard as indirect, in that it is a fine imposed on employers for creating employment. Apart from being counter-productive it is another onerous cost on the productive sector which is reflected in the price of goods and services to the consumer.

Direct Tax (Personal and Company)

Federal Income Tax was introduced in 1915 to help finance the First World War and Australians were, on an average, paying less than 6 per cent of their incomes in direct tax, even until the outbreak of the Second World War. It was presented as an emergency measure to cover the need of that time, but of course it was retained and is still with us half a century later. It was voted in under one guise but retained under another. Similarly, sales tax and payroll tax were introduced as temporary measures and were no doubt accepted without murmur by gullible taxpayers.

Income tax is of course a form of involuntary servitude as it means that all of us work a large part of the week (several days) for the government, before being allowed to enjoy the privilege of keeping the fruits of our own labour. (The essence of slavery, after all, is forced work for someone at little or no pay. Our existing system appears to meet this definition!)

Graduated (Progressive?) Tax Rates

The system of discriminatory, graduated taxation (accepted in Australia under the misleading name of progressive taxation, it could more accurately be called 'Regressive Taxation') is a form of expropriation of the fruits of the labour of all productive individuals. It is incompatible with the preservation of a market economy and was designed as a means of bringing about socialism. (Karl Marx always stressed that heavy 'progressive' income tax was a vital midwife of the socialist State.)

It can be said that the graduated income tax scale is an anti-work tax and prevents many people from being productive. In other words, 'the more you work, the more you work for nothing'. Another way of saying the same thing is that the more we work, the lower our hourly rate becomes. Such an anti-incentive tax system is hardly the way to direct our country toward productivity and prosperity.

Pay As You Earn (PAYE)

PAYE or Group Tax (called withholding tax in the U.S.A.) is a clearer cut instance of involuntary servitude. The employer is forced to expend time, labour and money in the process of deducting and transmitting his employees' taxes to the government – yet the employer is not recompensed for this considerable effort. What moral principle justifies the government forcing employers to act as its unpaid tax collectors?

The PAYE principle, of course, is the central weapon of the whole Federal Income Tax System. Without the steady and relatively painless process of deducting tax from the worker's pay cheque, our government could never hope to raise the high levels of tax from these same workers in one lump sum.

Once again, this dreadful 'weapon' was only introduced as a war-time expedient (this time for the Second World War), when we followed the U.S.A.'s lead. (Incidentally, we have a present-day, free-market advocate, Professor Milton Friedman, to thank for designing the mechanics of this system. He was working, at the time, as a junior bureaucrat in the U.S.A. Treasury Department (his wife Rose, also an economist, often jokingly states that she has still not forgiven him for participating in that project)).

Businessman, Miner and Farmer

In forcing business to become an army of unpaid tax collectors, the government has given business a weapon for their survival: down tools and refuse to co-operate. By doing this they could organise to throw a giant spanner in the works of the tax collecting machinery. This would force the government to retreat, as over 50 per cent of its revenue came from PAYE.

The government could not possibly cope with passive tax resistance on this scale. The repercussions of such concerted action would certainly put the case for the government doing its own 'dirty work' most effectively, instead of casting commerce in the role of 'unpaid thief'.

Over the past decade, I have studied the Australian mining community, business community and our farmers. Owing to the complex systems of protection and special favours bestowed by governments, unfortunately I can't see any possibility of these sectors leading a tax revolt.

Name me a business that doesn't have special favours of some kind from the government. They have a giant conflict of interest, as they cannot speak out for free enterprise while at the same time lobbying for special privileges and limits on competition.

The politicians and bureaucrats have tempted them to feast in the public trough with concessions of various kinds. They have been regulated to an extent where, at best, they have only a form of supplicant independence. Hardly enough independence to stand on their own two knees! This form of supplicant independence has led to the phrase 'the gutlessness of businesses'.

If you go into a Minister's waiting room, after any of the big free-enterprise seminars in Canberra, you will find that room full of important and earnest captains of industry, all looking rather embarrassed in the company they are in and all firmly clutching their small, flat, black briefcases containing their free enterprise halos. They are too shy to wear these while they are asking the government for another handout.

This lack of cussed independence throughout our business, mining and rural sector is unfortunate and it has created an ideal opportunity for our government to wheel in more amending legislation to tighten up and extend the PAYE system. If no resistance is met, we will see PAYE tax extended to partnerships, trusts, or to small companies for services provided under contracts. This will further weaken the base for any future constructive tax resistance.

Man was designed to walk upright but such continued cooperation between employers and the taxman lessons our chances of doing so.

The British Example

Ordinary, basic-wage-earning Britons now pay more to the taxman each week than they pay to the grocer.

How has this has happened in the U.K., when in 1939 they were an almost income-tax-free society (by today's standards), yet a society which provided most of the basic services and was also preparing for war?

Looking at taxes then and taxes now, one can't help but wonder: Where the hell has all the money gone? Of course in the U.K. one of the big differences is nationalisation. In 1939, privately owned electricity-generating companies, steel firms, coal mines and so on, were providing taxable income. They are today burdens upon, not aids to, the ordinary taxpayer. But even after allowing for all that, the abiding impression is one of waste and unnecessary state expenditure.

Subsidies in those days were more selective. Subsidised council houses, for example, were rarely provided for those perfectly capable of looking after themselves. The Britain of 1939 wouldn't have put up with the scale of welfare scrounging for which the honest wage earner has to pay these days.

Civil servants did not enjoy wages and conditions which made them the envy of the productive sector. Their current index-linked pension schemes, which grant them remarkable privileges, would hardly have been tolerated then.

However, perhaps the worst feature about 1978, as compared with 1939, is quite simply that massive taxation and massive government expenditure are taken for granted. An almost income-tax-free society, most believe, would mean unrepaired streets, negligible welfare, no defence and inadequate law and order (which overlooks the fact that many U.K. streets are neglected now, deserving cases often get little welfare, defence is inadequate and the Police Force in the U.K. is almost mutinous!) It is this acceptance of the inevitability of higher taxation which must be fought. It is remarkable, as the 1939 U.K. budget demonstrated, what a government can do if it had to.

The scale of waste and misuse of public funds escalate beyond control once the political parties are allowed to believe that the taxpayers will tolerate almost anything.

Australia – How Bad?

The foregoing U.K. example is relevant to our own concerns and it is always easier to talk more objectively about some other country.

It is quite an interesting and separate study, finding out just where the taxpayers' money is going. (Did you know that it costs taxpayers up to $1,500 each time a tattoo is removed from teenagers under the age of 18?)

The Taxpayers' Association is publishing much well-researched material in its magazine *The Taxpayer* and from examining these government expenditure charts it becomes so obvious that a considerable tax cut could be made without any reduction in essential services. Only the government frivolities need to be cut.

Not even the bureaucrats themselves could argue with the statement that 'Essential services will not necessarily be restricted if taxes are reduced'.

Let's look at some basic statistics here in Australia. Figures from the Institute of Public Affairs show the horrific facts in a simple table.

Taxation Burden – all levels of government – 1965 & 1976

	1965-66 ($m)	1976-77 ($m)	Increase per cent
Federal	4,221	19,770	368
State	589	3,060	419
Local	276	1,075	289
	5,086	**24,814**	**388**

In actual fact the real tax burden, on the Australian people, is even higher than reflected in the table above as Milton Friedman explains. "The real tax burden on the people is the amount the government spends, not the amount that it labels taxes".

That means if our Federal Government spends $29 billion in the financial year 1978-79 and takes in something like $26 billion in taxes, who do you suppose pays the other $3 billion deficit? Milton Friedman reminds us that it is not the 'tooth fairy', it is the taxpayer, through government-created inflation! Inflation is the cruellest tax of them all because it hits hardest at the prudent members of the older generation who have worked and saved all their lives to retain some independence in their later years.

Our Federal Government is spending $50,982 for every minute of the year. Nearly one in every three working Australians is a public servant and while Australia's population has increased an average 1.34 per cent each year since 1972. There is no way Federal and State

Government can justify the 38.7 per cent explosion in the number of civil servants that has occurred since 1972.

Total taxation – Federal, State and Local – now lifts $8,500 from the pockets of Australia's average family of two adults and two children, every year. This total accounts for well over 50 per cent of National Income. This means that the average Australian now works from 1 January to the end of July for one government or another before he begins to work for himself.

There is a growing awareness among Australians at all levels that they are being ripped off and the rumblings of revolt are growing louder. It is only a question of how bad things will have to become before Australians are moved to the realisation that what is happening is not inevitable.

Danger Point

Big government and high taxation either crush the spirit or raise rebellion. The economists Colin Clark and Lord Keynes stated in Britain in 1945 that things historically go wrong when taxation exceeds 25 per cent of National Income.

In Australia we have doubled this danger point and the best of our politicians are aiming to 'slow down the rate of increase'. As Australia has passed the historically calculated danger point, what is called for is a simple reduction, not a 'slowing of the rate of increase'.

Company Tax Even Worse

An August 1978 report on Company Tax by the Confederation of Australian Industry stated that many Australian companies are paying tax on real profits far in excess of the statutory rate of 46 per cent.

The Report stated that 50 companies included in the survey paid 96 per cent of their total real profits in tax. The results of this survey, which has been submitted to the Federal Treasurer, shows how Australian companies continue to be penalised by the present taxation system.

In the submission to the Treasurer it states that many companies have to borrow to pay tax and dividends. Others have to run down stocks. This affects their ability to meet, promptly, the demands of their customers and stops them from replacing obsolete plant. The result is "a most dangerous situation for the corporate sector in Australia today".

Our business leaders should be our main defenders of free enterprise, but they are not. In fact, they are often quite the opposite. This Confederation of Australian Industry (CAI) Report partly explains why.

Under this tax system, individuals and the corporate business sector have both lost economic freedom to the State where they are only about 30 per cent free. With the standard rate of company tax running at 46 per cent, the government effectively owns 46 per cent of every company in Australia. We talk of ourselves as a free enterprise society and yet in terms of the fundamental question about who owns the means of production, we are 46 per cent socialist, or even more as a result of the other factors taken into account by the CAI Report.

It has been said that the only difference in the degree of corporate socialism in Australia and communist Yugoslavia is exactly 20 per cent, as the Australian Government creams off 46 per cent of the profits as against the 66 per cent taken off in communist Yugoslavia.

Even these figures grossly understate the position of government control because of its effects in regulating business in areas other than taxation. That is why I feel we shouldn't wait for the lead to come from big business and the corporate sector.

Ordinary people and tax resistance

The free enterprise system based on the free market, private property, individual freedom/responsibility and limited constitutional government brought us many benefits which we still enjoy today. In creating the model for a non-coercive voluntary society, it exhibited a system that, whilst not perfect, at least minimised the faults of earlier systems.

If big business and organised groups, be they intellectuals, unions, etc. (all having been bought off, to varying degrees, by government favours), will not defend the free enterprise system and resist the continued growth of government, then this only leaves ordinary people, acting as individuals and groups, to mount some effective resistance. (No doubt we have a few equivalents to Thomas Payne, George Washington, Patrick Henry and Thomas Jefferson out there, to give some inspiration to all the others who want to act but don't know how.)

One encouraging sign is the growing visibility of government waste and inefficiency. Many people complain about government waste. They should welcome it, for two reasons.

In the first place, efficiency is not a desirable thing if somebody is doing a bad thing. Government is doing things we don't want it to do, so the more money it wastes the better.

In the second place, waste brings home to the public at large the fact that government is not an efficient and effective instrument for achieving its objectives. One of the greatest causes for hope is a growing disillusionment around the country with the idea that government is an all-wise, all-powerful, Big Brother who can solve every problem that comes along simply by throwing enough money at it.

In fact, if the government spends more, then the people have less to spend. Since the government spends money less efficiently than individuals spend their own money, then as government spending has gone up, the problems have become worse.

The continually repeated failures of government intervention in the economy have brought public to realise the undesirability of government intervention, resulting in the tide of public feeling that government is not the appropriate way to solve our problems.

How Did Government Get Control?

Why did we move from a situation in which we had an essentially free society to a situation of almost total government control?

Once again Milton Friedman puts it well when he states, "the fundamentals of most government intervention is an unholy coalition between well-meaning people seeking to do good on the one hand, and special interests (meaning you and me) on the other, taking advantage of those activities for their own purpose."

The general movement toward bigger government has not come about as a result of people with evil intentions trying to do evil. The great growth of government has come about because of good people trying to do well. But the method by which they have tried to do good has been basically flawed. They have tried to do well with other people's money.

Doing well with other people's money has two basic flaws: In the first place you never spend anyone else's money as carefully as you spend your own. So, a large fraction of the income is inevitably wasted. In the second place, and equally important, you cannot do well with other people's money unless you first get the money away from them. So that force (sending the taxman to pick someone's pocket) is fundamentally the philosophy of the welfare State. That is why the attempt by good people to do well has led to disastrous results.

If this is now becoming so obvious, why isn't everyone resisting?

I would suggest that resistance to taxation declines in some inverse proportion to the numbers who derive income from it.[11]

I don't know what the exact proportion is, but it sounds realistic to say that no-one can like taxation unless he is living off it. He may not even like it then, but he is certainly less likely to resist it.

Now the politicians with their misinterpretation of democracy (they neglect the rights of individuals and minority groups), see democracy as unbridled 'mob rule' and feel that if they have the numbers, anything goes. They see the majority of Australians as being in favour of taxation because they are claimants on the State and it is these politicians who also make clever use of the media to create two illusions:

- That it is always someone else that is paying the tax and never the groups to which they are directing their election promises.

- That taxes are only necessary because Australians have demanded more public services – particularly health, education, transport and welfare services.

The public is only demanding these 'unlimited goodies' because they have been promised and offered by the politicians as being 'free', in their attempts to retain their political power. Their assumption – that the majority of Australians are in some way claimants on the State – is probably correct. For a start, almost 30 per cent of the population work for one of the various arms of the government. If you add to this percentage all those in receipt of pensions, unemployment benefits, government grants, and government educational funding, then it is pretty obvious that well over 50 per cent of the population SHOULD be in favour of taxation.

Therefore, in the minds of politicians any tax resistance will be

---

[11] I would suggest that resistance to taxation declines in some inverse proportion to the numbers who derive incomes from it. A recent report from KPMG found that "60 per cent of households pay the same or lower amounts of income tax compared to the payments they receive from the government".

isolated and of no great concern to them.

What they overlook are two vital points:

- The size of the tax contribution made by productive people is now reaching the stage where it can no longer be concealed by more election promises.

  A stark comparison of government tax/tariff costs was recently pointed out to the cattle industry.

  ➢ In Japan 2 bullocks will buy a Toyota Landcruiser (cab and chassis).

  ➢ In the U.K. 3 bullocks will buy the same vehicle.

  ➢ In Australia 31 bullocks are needed – 10 times the input price of one of our major trading partners.

- The other point overlooked by the government is that a large percentage of what they consider to be tax claimants can no longer, in all conscience, identify with the disastrous drift toward the welfare state and that many of these same people have sufficient confidence in their own ability to realise that they would 'make it', quite often with less red tape and more job satisfaction, when working for a free-enterprise employer rather than the government.

  I can't find many people who are working for the government who are prepared to proudly identify themselves with the current drift and the burden on them is becoming unbearable. This is highlighted by some recently released figures showing the increase in the proportion of our population now living substantially on welfare. In 1972 it was 9 per cent and now it is 15 per cent (excluding repatriation pensions, family allowances, etc.). This is doubly serious for all of us because every extra tax dependant is one less tax payer – so it is a double drain on the system. [The figure for working-age Australians who are on welfare is still around 15 per cent.]

I haven't painted the picture any bleaker than it really is. The only good news that I could venture is that it is almost impossible for the recent percentage rates of increase to continue. This would mean that our lovely government would be spending something like 200 per cent of the country's National Income in the late 1980s. [Current spending from Commonwealth, state, and local government is around 37 per cent of GDP. You could make reference to debt, which is of course how governments can increase spending without raising current taxes. Over the last decade, spending by the Commonwealth government alone has risen by 70 per cent and gross government debt has risen by around 870 per cent, or around half a trillion dollars. While the population has grown by just 18 per cent over that period.]

It does become clear, however, that, without a tax revolt in one form or another, the situation will continue to get worse each year as we witness the process of democratic degeneration.

> Government can't give us anything without
> depriving us of something else.
> – Henry Hazlitt in article "Instead of What?" – FEE.org

> Elections; are advance auction sales of stolen goods.– H.L.
> Mencken

The Income Tax people have streamlined their tax forms for this year. They go like this:

- How much did you make last year?
- How much have you left?
- Send it!

A new restaurant opened under the name of The Taxman. The Proprietor explained that he felt the title to be appropriate as the Taxman would enjoy most of the profits.

Encourage government waste? Yes! If the government is 'stuffing up' our country. Do you want them to do it more efficiently?

> The world is ruled by letting things take their course. It cannot be
> ruled by interfering. – Lao Tsu, 6th Century B.C.

Recognising the enemy and studying his bad manners

"The Australian Taxation Department (hereinafter referred to as the Tax Department) spent $60,000 of your money, in 1977, on hired pot plants, as they provide an 'aesthetically pleasing' work environment!"

By now I feel that many of us are coming to realise that the trials of the '70s are the effects – primary and secondary – of government policies.

This makes the government, no matter how well-intentioned, the enemy. Their front-line storm troopers are the 12,000 strong battalion called the Tax Department.[12]

Incidentally, in 1966 you gave these people $3.3 million extra in overtime payments so that they could persecute you more effectively. The figure increased to $3.5 million in 1977 and with the present government's paranoia about blocking all the so-called loop-holes,

---

[12] This makes the government (no matter how well-intentioned) the enemy, and their front-line storm troopers are the 18,234 strong battalion called the Tax Office.

it is not unlikely that the figure will double this year.[13]

Even with their 12,000 strong battalion and their dubious behaviour, it never ceases to amaze me that they have carefully concocted some sort of 'Santa Claus' illusion where many people hold them in high esteem. The only time those people hear from the taxman is when their annual tax refund cheque is received through the mail.

Until the employers cease their role as unpaid tax collectors (PAYE) and force the ATO to directly confront all Australians, this Santa Claus illusion will persist.

Nothing personal!

On legal advice, I am refraining from covering any aspects of my own experiences with the Tax Department as it is not in my rational self-interest to invite reprisals.

They are a fairly humourless bunch and I can't help recalling the riddle: Why is the Tax Department like a septic tank? Answer: Because the big pieces always rise to the top!

Bad behaviour

Let me give you some examples of the bad behaviour you can expect from Tax Department personnel (it is interesting to note that these people are euphemistically called 'public servants').

From the outset it is important to appreciate the reason for their self-righteous insistence that they are only carrying out 'the law'. This should surprise no-one, as all tyrants use the law quite unscrupulously: Allan Bullock, the historian, tells us, "Hitler

---

[13]   Under their enterprise agreement, ATO staff are only required to work until 4:51pm each day. A proposal by the government in 2017 to increase the work day by nine minutes – to 5:00pm – was met with backlash from ATO staff until the proposal was dumped.

recognised the enormous psychological value of having the law on his side. He never abandoned the cloak of legality."

This 'law', known as the *Income Tax Assessment Act*, has been described as 'An abomination of intrication'. Before handing down his judgement in four tax cases in 1976, Mr. Justice McInerney made the following statement, "The income tax appeals, in respect of which I am giving judgement this morning, have involved consideration of various provisions of the Income Assessment Act, the drafting of which is so involved and tortuous as almost to defy comprehension. In a loose paraphrase of the words of the prophet Daniel, the Act could be described as an abomination of intrication."

Armed on one hand with the Tax Act, and on the other hand with skilled techniques in the use of fear and coercion, the Tax Department has created an illusion where members of the general public believe the Tax Department to be virtually omnipotent. The agents have the power, it is believed, to come into the home or office to seize books and records, levy, seize and sell assets without due process of law and all on the basis of some obscure administrative decision which took place in the back room of one of their supervisors' offices.

From the ordinary taxpayer's position, however, this is simply not true (in 1978). It is 90 per cent bluff. If the general public knew more about the limitations on the Tax Department's power, the individual would be better able to protect himself against over-reaching Tax Department agents. Tax Department staff are trained and expected to take advantage of the lack of knowledge the ordinary citizen has of his individual rights.

If most of our citizens don't even know what their rights are, how can they expect the Government to protect these rights? It is only the skilled techniques of these Tax Department investigators that allow them to get away with many of their dubious actions. If the taxpayers in this country ever discover that our Tax Department operates on 90 per cent bluff, their entire system will collapse.

My good friend, and noted Californian tax attorney, Clyde R. Maxwell, made some interesting observations along these lines which he published under the self-explanatory Tax Department title "Powers which the Tax Department does not have but which the general public thinks it has".

Another excellent article by Peter Clyne, "What to do when the taxman calls on you" explains how "most tax investigations would fail dismally if taxpayers knew about and exercised their right of silence" and gives examples of how citizens are often bluffed into cooperation.

One case of which I heard recently, involved a businessman being called into the Tax Department where they stated that he owed them $27,000. He disputed this but they insisted that he owed it and they demanded payment. He went to his accountants and they fought the case for a year. At the end of the year the Tax Department admitted that he owed them only $11,000. His accountants advised that this amount be paid and that they keep on fighting. They fought for another two years and finally the Tax Department agreed that he hadn't owed them anything. He got his cheque for the $11,000. The thing is, that if he hadn't fought them, they would have collected $27,000 and that would have been considered just taxes.

So many people don't fight them. It is extortion. Of course, often you can't afford to fight them. One thing that is obvious is that it is a most unequal battle. The Tax Department can keep on spending money forever, because they can tax all the people in Australia to pay these bills. The taxpayer has nobody to tax. What happens to the poor taxpayer who doesn't have any money? He is 'dead' unless he indulges in some 'tax resistance on a limited budget' (see later in this chapter). Further, if a citizen is in trouble with the Tax Department, they usually seize all his money, leaving him without anything to finance his defence.

In Western Australia the Tax Department, with ruthless precision, moved 16 tax investigators in on one businessman. The operation was carried out with considerable skill. Four investigators on the businessman, four on his son, four at his accountant's office and four at his bank. After considerable dislocation to his business, his accountant and bank manager, the sixteen investigators pulled out simultaneously without having found any irregularities whatsoever. It is unfortunate that they weren't even well mannered enough to apologise for the inconvenience caused.

It should always be remembered that the Tax Commissioner has been granted wide powers to obtain any information he requires in order to make an assessment on any particular taxpayer. He may inspect all the taxpayer's records and those of any other person if he believes the relevant information is there. When you think about it, this is a very far-reaching power but there are limits to the manner in which he obtains this information.

Many taxpayers believe the Commissioner can only go back six years from the date upon which the tax under the relevant assessment became due and payable. But it is most important to note that, if the Commissioner, or his henchmen, suspect a tax liability arising from fraud or evasion by the taxpayer, then they can go back and amend the relevant assessment at any time.

Contrary to popular belief, however, the Tax Act does not clearly distinguish between tax avoidance (arranging your affairs to minimise tax) and tax evasion (cheating). Telling lies (non-disclosure or incorrect disclosure) is of course an offence under the Act and of course is not recommended. Avoision, is a relatively new word for the arrangement of one's financial affairs so as to avoid or minimize tax liability in such a way that it is not clear whether this constitutes lawful avoidance of tax or illegal tax evasion.

Section 260 of the Act seems to give the Tax Department sweeping powers to over-ride any arrangements designed to reduce the burden of tax. But the past few years have seen the High Court interpret the Act in favour of the taxpayers. This is reasonable as the taxman should not benefit from any law that is in anyway ambiguous. As Clyde R Maxwell commented: "In any civilised country, where the wording of any law is ambiguous, then it must be interpreted in favour of the citizen, never in favour of those who drafted the law."

Savage treatment

A taxpayer paid his tax assessment in full, then 18 months later got a further bill from the Tax Department. He couldn't believe it. Not only had the Tax Department made a mistake but they billed him for the 18 month's interest on the mistake. He had to borrow the money from the Bank and pay interest on that. Then, in order to pay the Bank, he had to cash in an insurance policy, which he had been paying for some time for one of his children. He is now a life-long enemy of the Tax Department.

Other sundry stories are typical of the brutality dispensed by the Tax Department and elsewhere. Their computers are not programmed to dispense sympathy.

One man committed suicide, leaving a note to explain that he couldn't endure any more harassment from the Tax Department. His 17-year-old son, who had started work for the first time, was due a $400 tax refund. Instead of the refund, he received a notice from the Tax Department stating that his $400 had been applied to his dead father's tax.

Another businessman's desk was searched by a tax investigator. Finding a letter that seemed to imply a romance with "another woman", the investigator rushed to show a copy to the victim's wife. Apparently, he hoped that he could persuade the woman to inform against her husband.

Endless examples are available to show how the Tax Department has the power to bankrupt any citizen, causing permanent disruption, even if, eventually, the taxpayer is proved innocent.

The Tax Department's methods include picking locks, stealing records, illegally tapping phones, intercepting and reading personal mail, using hidden microphones to eavesdrop on private conversation, using undercover agents with assumed identities, sexual entrapment and other assorted crimes.

Some tax investigators justify their actions in violating these laws by stating that it is part of their 'duty' in administering the Tax Act.

In Western Australia many of us have had instances of undercover tax investigators hanging around hotel bars and randomly striking up conversation with customers, hoping to piece together various aspects of mineral claim transactions.

This raises an interesting point on ethics. We know it is wrong to initiate behaviour of this kind, but it is permissible to retaliate in kind.

This may have stimulated a well-known taxpayer, who had been constantly pestered by the Tax Department, to repay the investigator 'like-for-like', in turn, making his life unpleasant. The taxpayer hired a private investigator and got a run down on the taxman's previous employment, which revealed that he had failed at several businesses. He then sent an interviewer to question the taxman's previous employers, associates and neighbours. The investigation extended to contact with former school mates. This investigation also unearthed a collection of 'bad habits'. Compiling a history of the flops and failures in the taxman's life, the taxpayer then had copies distributed to all letterboxes in the taxman's suburb. The obvious theme of the piece was that the taxman had been unable to make an honest living and now he was working with the government, trying to bully people into parting with their hard-earned money. The effects on this taxman and his family were somewhat unsettling. I leave it to you to pass judgement on such retaliatory action.

How safe is money in the bank?

Press headlines – "Tax Grab From Bank Account" (*Sunday Times*, 13 March 1977) drew attention to the fact that in Australia (as in several East European countries) the government is empowered to draw money from your bank account if it 'thinks' that money is owing for tax (Section 218 of the Income Tax Act). The interesting thing about the case in question was that it had not even been proved that the person actually owed the tax. The case was not to be heard until six weeks later but that did not deter the Tax Department's 'tax grab'. This is an interesting thought for anyone who leaves money lying around in banks, imagining that it is safe!

Loopholes

The taxman's definition of a loophole is 'any provision that allows an individual to keep any of his property or money'.

The present government's blitz on so-called loopholes, as well as being futile (some of the best brains in Australia are devising new tax avoidance schemes at a faster rate than they can be legislated against), is extremely dangerous, as it will lead to an expansion of the Tax Act, already regarded as an 'abomination of intrication'. A future government would be delighted to discover that no further legislation is required as they already have total control over our economic freedoms.

As long as a country such as Australia countenances a bureaucracy which can terrorise the public like a mad dog, and victimise all those who fail to fight, that country should recognise that it has the machinery all ready for an authoritarian government. If this ever happens, they will find the machinery of oppression ready oiled and in good working condition.

The root of this evil is the seizure of property without due legal process. Once again, their actions are 90 per cent bluff, and the myth is perpetuated by some bystanders who say, 'Well, the taxman has to be able to collect his dues!'

How do other people collect their just debts? If I think you owe me money, I argue in court, and if I get a judgement against you, you have to pay. The whole business community of the nation, the whole population, is able to get by with this due process of law.

If the Tax Department wants money from you, they don't even have to have a good reason. All they do is say, "You owe us money. We seize your property. When you prove that you don't owe us money, we will give back your property."

This 'vice' promises to destroy all our freedom.

No living person in Australia has the right to seize property merely because they claim debt. However, this mindless bureaucracy behaves as though it has that right. They will continue unchecked on this stampede unless they are constantly challenged by concerned taxpayers.

The Tax Department would do well to heed the advice of Lao Tsu: "Do not intrude in their homes. Do not harass them at work. If you do not interfere, they will not weary of you."

- Bureaucracy – A perpetual inertia machine. The majority party that never loses an election. The rulers without faces.
- The Free World – Those nations not captured by communists. i.e. Those nations captured by non-communists.
- A little boy very badly wanted $100 and is mother told him to pray to God for it. He prayed and prayed but for two weeks nothing happened.

  Then he decided that perhaps he should write a letter to God requesting the $100. When postal authorities received the letter addressed to God they opened it, read it and decided to send it to the Prime Minister.

The Prime Minister was so touched and aroused that he instructed his secretary to send the little boy a cheque for $5. The Prime Minister thought this would appear to be a lot of money for a little boy.

Upon receiving the $5 the little boy was delighted and immediately sat down to write a thank you letter to God which went as follows:-

"Dear God,

Thank you very much for sending me the money. I noticed that you had to send it through Canberra. As usual, those bastards deducted 95 per cent."

Tax Revolt or Tax Resistance?

If you don't do anything, you are guilty of compounding the problems generated by the Welfare State. By voluntarily going along with all the Tax Department's claims you could share in the guilt of 'aiding and abetting criminals'.

At present levels of government spending, no rational tax system is possible. However, if Government spending can be reduced to a reasonable proportion of the national income (even 15 per cent), then, and only then, can some sensible moves be made toward cutting taxation back to a level where incentives can encourage work, production and income.

Politicians and bureaucrats are generally short on realism, in that they don't grasp the fact that if you reduce the after-tax reward for something, you get less of it and if you increase the subsidy for something, you get more of it.

In Australia, today, we are taxing work, savings, investment, thrift, productivity, effort, success and risk and we subsidise non-work, unemployment, debt, borrowing, consumption, leisure, idleness and mediocrity and we are getting much more of the latter than we are of the former.

Though it sounds simple, unfortunately nothing much will change until enough taxpayers become aware of the crippling effects of excessive government intervention in our daily lives and mount their own Tax Resistance Programs.

If this happens, our politicians will be forced to take steps to limit government activity to its correct role of referee and protector of individual's rights and their property.

One example of how Australia's incentives to work have been destroyed is reflected in a recent letter to the *Bulletin*:

Taxed into Idleness

… When I started in my present business as a dental mechanic, 10 years ago, I figured I needed to work an 8½ hr day for four days of the week and an 11½ hr day on the remaining day and this worked well until I saw half my energy and none of my frustrations were being eaten up by the Tax Department.

Now I work seven hours per day, four days per week. The fifth day, Friday, I sit about, do my own thing – I even have time to write to the *Bulletin*. Life – be in it.

– R. Johnston, Hobart, Tasmania

The wrong people are the worst affected by our crippling tax structure. As soon as a single aged pensioner, under 70, gets more than $28 per week over his pension, he loses up to 66¾ per cent of each extra dollar income. While this slug is not officially called a tax, it is still a treasury grab and a grab by any other name is still a tax. The fact now emerges that pensioners under 70 with a tiny private income now effectively pay the highest rate of tax in Australia. Our lovely government now plans to go further and has indicated its intention to extend this arrangement to all old-age pensioners, even those over 70.

Our over-regulated Welfare Sate encourages an unprecedented degree of financial evasiveness, made necessary for its citizens' economic survival. A recent comment from a bank manager in neighbouring New Zealand stated that because of the enormous proliferation of regulations in every form of commercial endeavour, intelligent Kiwis spent at least 25 per cent of their time working out ways to get around those rules.

This sounds familiar. We are in the same area, for the same reason. Ordinary people are making their own small moves to protect themselves. One such move is the enormous growth of dishonesty in all matters surround taxation and social security. This movement will grow and swell into an increasing expression of civil disobedience. It is to be applauded!

Ordinary people feel lost and are trying to get back some control over their own money and lives. The growth of the 'subterranean economy' is one such example. More and more of the income and transactions in the country go unreported. More and more people are working in ways where their earnings will not be written down. This goes right across the whole range of our society. Pensioners, who are restricted in their earning rights, resort to a wide range of ruses to increase their incomes without endangering their pensions. Armies of lawyers are devoting their energies to cheating on taxes for their clients. When they succeed, they are treated as heroes. Fear of the tax collectors is gradually turning to hatred.

There is a premium to be earned by opting out of the normal economy and going into the 'subterranean economy'. Taxes can be avoided. Cash transactions – even harder transactions – are much better than the old style of business with its burdensome bookkeeping. Again, beware the fine line the Act draws between avoidance on the one hand and evasion and telling lies on the other and avoid the latter.

In many countries the level of 'subterranean' economic activity is nearing 20 per cent and I heard recently that in India they are unable to complete their nationally published economic figures without reference to 'unofficial trade figures'. The City University of New York estimates the USA's underground economy at US$150 billion or about 10 per cent of the Gross National Product. The Inland Revenue estimates UK's figures to be 7½ per cent of GNP, so it appears reasonable to expect Australia's figure to be around 7½ per cent to 10 per cent.

The average citizen sees this moral degeneration taking place and would like to do something, but usually feels powerless. The choice is between action and inaction. Inaction is reminiscent of the words of Edmund Burke, "It is only necessary for enough good men to do nothing, for the forces of evil to win."

Several options of action are now open and they are popularly described as tax revolt and tax resistance. Tax revolt appears to be a stronger term, involving visions of violence (or at least civil disobedience), usually giving the impression that others are involved as a result of their circumstances being identical. Tax revolt is probably the high-profile species of tax resistance. Its main objective it to publicise the evils of the tax system and thereby undermine it. Only those who like to live dangerously should participate in a really high-profile tax revolt; it is good for the adrenaline but bad for the nerves. Just how high-profile the activities of a person should go is a judgement each person has to make for himself.

Tax resistance, on the other hand, involves a much broader range of techniques which can be applied by every single taxpayer to the extent of his personal requirements. It is a fun game that can be played by any number of consenting adults. It comes in two common forms:

Sneaky tax resistance which involves various devices which are usually illegal but keep one's money away from the tax collector. By definition, sneaky tax resistance is illegal and if caught at it, one will probably be convicted. This approach is not recommended, simply because there are alternatives which result in minimal taxation with little or no risk of criminal prosecution.

Low-profile legitimate tax resistance involves using the tax laws legally in order to minimise the tax burden, as much as possible, at relatively low risk. It also involves knowing your rights and avoiding any attempted con tricks from the Tax Commissioner.

The first rule for successful tax resistance is to always tell the truth (i.e., if you have to say anything at all!). No-one has ever gone to jail in Australia for not paying tax, but they have gone to jail for telling lies.

There is more challenge and job satisfaction in finding your way around the 'abomination of intricacies' rather than simply breaking laws (even more skill is needed to arrange for bad laws to be repealed altogether).

Tax resistance is about the only sensible way of economic survival for productive people in these days of increasing government and tax burdens. It is becoming increasingly obvious that the only way to cut government spending is not to give them the money in the first place. As Lao Tsu remarked: "When the country is ruled with a light hand, the people are simple. When the country is ruled with severity, the people are cunning."

- Taxpayers are the only people who don't have to sit through an interview to work for the government. (Perhaps they are the only people who actually work for the government?)
- The successful tax resister is one who commands such respect that when he phones the Tax Department (collect – reverse charges) he is told, "I am sorry they are all out!" What, all 12,000 of them?
- You can't free slaves. They must free themselves.

Economic Reasons for Tax Resistance

The Tax Department has lifted more money from the purses, wallets, socks, mattresses and bank accounts of the Australian public than the combined efforts of all other thieves in the whole history of Australia. The great power which the Australian Government (and its servants the Tax Department) has, over the years, arrogated to itself has been constantly misused with dangerous results. Today every productive citizen, no matter what income bracket, is a direct economic victim of these policies.

As a result, the entire middle class which is the basis of stability and prosperity of the nation is being steadily destroyed at an accelerated rate, not only by irrational, unconstitutional confiscatory taxation but by the wholesale destruction of Australia's monetary unit, the dollar, by irresponsible government counterfeiting, by printing money without asset backing and by over-regulation of mining, agriculture and other industries, thus preventing them from achieving reasonable performance.

What about the unemployed?

The politician's idea of helping the unemployed or poor to become helpless is no act of kindness. The greatest threat to our unemployed and poor is the slow rate of growth of our economy.

Successful tax resistance leaves more funds in the hands of individuals and industry, increasing investment and expansion, thus expanding the economy and in turn the tax base. Tax receivals would remain about the same but at least those who want to work could have a job.

Many people develop a mental block about free enterprise because it has the ability to make rich people richer. They overlook its far greater potential for making poor people wealthy.

Revolt and resistance, for economic reasons, should go beyond tax and include over-regulation of commerce and trade.

It was no accident that the industrial revolution took place in the late eighteenth century Great Britain. It was a period of minimum government regulation. As a matter of fact, the industrial revolution, one of the most important events in human history, seems to have occurred without the British Government even noticing. By the time they did, it was too late. Happily, otherwise they would probably have stopped it. (Sounds like Australia's mining boom.)

Adam Smith, 200 years ago, recognised that the free enterprise economy, not government-protected corporate capitalism and big government, are natural and probably irreconcilable enemies. He lived at a time when individuals, trade and commerce were bogged down in a morass of taxes, rules and regulations, introduced for the benefit of a few big business interests.

Adam Smith came to the radical conclusion that the wealth of the nation would be maximised if the government got right out of the way and limited itself to acting merely as a referee protector and peace-keeper. This radical idea was to change the world. It guided the development of the United States and ushered in the industrial revolution which brought the greatest increase in wealth and population that the world has ever seen.

In the past century we have come full circle with rules, regulations, restraints and taxes increasing daily. We need the ideas of Adam Smith and other free market economists to act as a catalyst in forging a coalition between Australia's intellectuals and our workforce to turn the tide against every growing government intervention into the economy and our daily lives.

Our independence has been illegally reduced

The simplest, easiest and most effective method of increasing our independence is tax revolt and tax resistance, the refusal to pay for our own destruction. It is the only one power each and every one of us possesses individually. Proposing to keep more of our own money is not asking for a free lunch. In fact, it is suggesting that people keep enough of their own money to buy their own lunch.

Moral Reasons for Tax Resistance

> No man in this country is under the smallest obligation, moral or other, so to arrange his legal relations to his business or to his property as to enable the Tax Department to put the largest possible shovel into his stores. The Tax Department is not slow and quite rightly – to take every advantage which is open to it under the taxing statutes for the purpose of depleting the taxpayer's pocket. And the taxpayer is, in like manner, entitled to be astute to prevent, so far as he honestly can, the depletion of his means by the Taxation Department.
>
> – Lord Clyde, Ayrshire Pullman Motor Case (1929) – 14TC754

Morality must be brought into this discussion as taxation, plunder or expropriation of the earnings and property of individuals is a moral question, as well as an economic one. It involves a challenge for the leaders of all free or semi-free countries such as Australia.

The challenge goes to all leaders, whether academics or industrial, unions or politicians.

Unless we restore the moral will to be independent, we will continue to sink into slavery and discussions about details of tax reforms will remain purely academic. To adopt a moral position with respect to tax resistance it is necessary to come to grips with a few concepts like the proper function of the law, the rule of law and the balance between a government administering the law or overriding it.

Frédéric Bastiat, the French economist and statesman, asked "If anything more than the absence of plunder could be required of the law?" Can the law which necessarily requires the use of force rationally be used for anything except protecting the rights of everyone? Bastiat suggested that for anyone to extend the law beyond this purpose would be perverting it and turning might against right. He maintained that the true solution, so long searched for in the area of social relationships, is contained in these simple words: "law is organized justice".

In France, in 1850, Bastiat predicted how socialist politicians would seize the opportunity of buying popularity and votes from the masses of people who want to consume something they have not earned. By perverting the law, these politicians could extend the definition of the law, to not only be just but also to be philanthropic. They could also extend the law to directly provide welfare, education and morality throughout the nation.

The fallacy of this extended definition of the law was that "everybody wishes to live at the expense of the State, but they forget that the State lives at the expense of everybody". When you start on this road to gain popularity by doing 'good' with other people's money it is easy at first. You have got a lot of people to pay taxes and a small number of people with whom you are trying to do 'good'. However, the later stages become harder and harder. As the number of people on the receiving end grows, you end up in the position where you are taxing 50 per cent of the people to help 50 per cent of the people. Or, 100 per cent of the people, to distribute the benefits to 100 per cent. The main problem, of course, is that the government absorbs most of the money on the way through.

Bastiat described this concept, of living at the expense of others, as "the seductive lure of socialism" and pointed out that these two uses of the law, i.e., justice and welfare, are in direct contradiction to each other.

We must choose between them. A citizen cannot, at the same time, be free and not free.

I would venture to suggest that nothing has contributed more to the loss of freedom, liberty and rights than the failure of the individual citizen to understand what the term 'rule of law' really means. Nobel Prize-winning economist F.A. Hayek argues that the 'rule of law' means that the coercive power of government is limited by rules pre-established. He states "Stripped of all technicalities, this means that government in all its actions is bound by rules fixed and announced beforehand …"[14]

It excludes all arbitrariness. It means that all men are equal before the law. It means that those who govern are not above the law and that the inalienable rights of the individual are supreme, above and superior to government. It means that the only object of government is, "to protect the rights of the individual; that the power of those who govern is delegated to them by the people; that the people, collectively, cannot delegate a power to government that they do not possess individually, i.e., people cannot delegate a power to government to do evil". It is impossible for the people to delegate a power to government to invade the rights of individuals, no matter how great the majority is that seeks to do so.

---

[14]   F.A. Hayek, *Road to Serfdom*, p. 72.

Not many of our politicians have come to grips with the true meaning of the rule of law or limited government. They are like backyard brain surgeons in the way they consistently confuse limited constitutional democracy with 'mob rule', where anything goes as long as you have got the biggest gang. Governments that behave the way ours do come under condemnation from none other than St. Augustine, "Remota Justitia quid sunt regna nisi magna latrocinia", which translates as, "A government which has departed from justice is nothing but large scale gangsterism." At this point I don't think that questioning some of our politicians' laws can be avoided, especially their application of the Tax Act.

It is only by recognising the conflict between a given rule and the rest of our moral beliefs that we can justify our rejection of an established rule. Even the success of an innovation by a rule-breaker and the trust of those who follow him has to be bought by the esteem he had earned by the scrupulous observation of most of the existing rules. To become legitimised, the new rules have to obtain the approval of society at large, not by a formal vote but by gradually spreading acceptance.

The very immorality of many of our government's actions in seeking and dispensing favours, in my view lays a very strong case for responsible citizens to join the Tax Resistance Movement on strong moral grounds.

- What country can preserve its liberties, if its rulers are not warned from time to time, that the people preserve the spirit of resistance? – Thomas Jefferson.

- It is not a man's duty, as a matter of course, to devote himself to the eradication of any, even the most enormous wrong; he may still properly have other concerns to engage him but it is his duty, at least, to wash his hand of it and if he gives it no thought longer, not to give it practically his support. – Henry Thoreau, 1948.

- Beware of strong drink. It could make you shoot at tax collectors, and miss. – Robert Heinlein.

Tax Resistance on a Limited Budget

Why are the people starving?

Because the rulers eat up all the money in taxes.

Therefore, the people are starving.

Why are the people rebellious?

Because the rulers interfere too much.

Therefore, they are rebellious.

– Lao Tsu, 6th Century BC

How can you possibly afford the legal costs involved in fighting the Tax Department? Most of the people I know who have tackled the Tax Department have usually had a great deal of no-charge legal assistance for two main reasons:

a. The legal profession is prevented from advertising, but if a lawyer has a win against the taxman, it is regarded much the same way as St. George's slaying of the dragon and consequently results in much good publicity for him.

b. Top legal professionals usually enjoy any real opportunity to call the taxman's bluff.

Someone recently said to me (but I don't know if they were joking) that "The kindest thing I could have ever done for my children would be not to have registered them at birth. They wouldn't have to put up with all this taxation nonsense."

Some tips and stratagems

Barter, smuggling and active participation in the subterranean economy.

Know your rights. Know what information the Tax Department can demand and don't provide one extra fragment of information.

Never cooperate with taxation investigators in any way. You shouldn't be expected to aid a thief when he walks into your home and, likewise, you shouldn't have to take him by the hand and lead him to the drawer in which you store your valuables. Any efforts you make in cooperating with them comes close to cooperating with suspected criminals. They are the ones who should feel guilty, not you. After all, whose money is it?

There may be merit in starting your own church. It doesn't matter how small the church is, you may be the only member, but as long as you hold regular meetings you may be able to enjoy all manner of tax benefits. In Australia the government guarantees freedom of religion and if they take any discriminatory action against your church, you would want to know why!

Stall, stall and stall. If you are investigated and you receive one of the Tax Department's long itemised questionnaires, delay replying as long as possible. Then, when they run out of patience, ask for a copy of their original letter to you as you appear to have mislaid it. When you reply, with very exact answers, give the wrong reference numbers and file numbers. It could be that they will be unable to match up and will be too embarrassed to pursue the matter. Eventually, they will write, asking for more detailed answers, in the hope that you will give more clues to their original questions, enabling them to trace the reference numbers you quoted. So far, you have gained a delay of about 18 months, without really disclosing any information. When you reply, this time, quoting the correct reference numbers, you should give exact answers but to entirely different questions (fictitious questions). This will start the cycle over again and you can bombard them with requests for a copy of their original letter, which they never ever wrote.

The challenge is to keep this sort of thing going for as long as possible as it pre-occupies their staff and keeps them out of your hair.

Don't try to prove your gross income. Just provide for your expenses. Keep all receipts in a large garbage can, this will build up after a few years. If you are investigated, don't let them take these receipts away, let them inspect the receipts on your premises and give them a good 'stir-up' every day after they leave.

When you are investigated, among the first things they do is run to your bank without your knowledge and photocopy bank records, cancelled cheques, etc. It may be best to have your bank cheque forms printed on dark red paper, if possible, as it is difficult to distinguish from ink when microfilmed or photocopied. There is also available a particularly fine pen, called 'commercial light blue illustrator non reproducing pen', that will not reproduce on microfilm copies or photocopies. The object of this type of activity is linked to the taxman having to generate a certain dollar volume for each hour of work. It is possible that if you can run up their time, they may wish to avoid you. With a bit of luck, they may even take you off their mailing list completely!

A sure-fire way of upsetting them appears to be including some unspecified 'sundry' income even if it is only a few hundred dollars. They keep insisting that you inform them where this income was generated. They never seem to believe your statement that you must be honest, otherwise you wouldn't have declared it and it is really the undeclared income that should concern them.

Remember: never volunteer any information.

Get yourself an accountant whose only loyalty is to you and not the Tax Department. This may sound strange, but you should bear in mind that it is the 'enemy' that issues your accountant with the licence that permits his livelihood. Some accountants express discomfort at the conflict of interest between complete loyalty to their client and clinging to their government-issued licence which may be withdrawn at the arbitrary whim of a bureaucrat. Many accountants now regret falling for the 'licencing trick' presented to them as a way of restricting competition in their field. It may now be necessary to go outside the conventional accounting profession for real protection. This may explain why most tax minimisation schemes are conceived by others and merely implemented by the accountants.

As a general rule, you should bear in mind that your records at accountant's offices and banks are on show to the taxman, usually without any notification to you, but at the present time records held by your solicitor are not so readily available.

I know of a man who had a small company and for some reason the Tax Department was pushing for a large tax payment. The company proprietor couldn't understand why he was expected to do all the work and have the Tax Department cream off all the profits. So, as the company had no assets, he neatly packaged the company books, including written letters of resignation from the directors and secretary, tied the whole lot in pink ribbon and mailed the lot off to the Tax Department – without stamps, of course! His covering letter, to the Tax Department, indicated that if they expected to get the profits, then he expected them to do the work. That is the last he ever heard of that matter. (Never waste your cash by putting postage stamps on letters to the Tax Department. They will always accept the letters in the hope that you might be sending them money!)

Use their own techniques. It is very difficult to obtain an opinion from the Tax Department. If you write, asking if tax will be payable on a certain venture, they usually write back suggesting that you 'try it and see'. In other words, they never wish to give an opinion that is binding.

Information you submit to them should be provided on a similar basis. You should cover yourself by writing 'without prejudice' on any letters, documents or tax returns. On that basis your return may not be used as evidence against you. You could even go further by typing on them 'The statements on this form were not made voluntarily but were extracted from me under threat of penalty of law.'

When replying to their correspondence, always try to establish some good human relationship with your investigator. Apart from asking about their family and ancestry make the odd enquiry about just what they are doing with all the money they are ripping off their fellow Australians. Ask the question, from time to time, as to just why they don't go out and get a 'proper' job.

Never answer any questions until the questioning bureaucrat completes and signs a copy of the 'Questionnaire for Bureaucrats'. By right you are entitled to know who your accuser is and the exact nature of the investigation. Usually the bureaucrats retreat rather than sign such a document.

Avoid phone discussions and endeavour to do everything by mail.

Make sure, when you are sending photocopies of anything

requested, that they are almost impossible to read. When they ask for better copies ask them to send back their copy so you can photocopy it better this time because you destroyed the original.

If, eventually, you are forced into a personal interview, make sure you have your lawyer, accountant and tape recorder with you.

Buy and read Harry Browne's book *Complete Guide to Swiss Banks*.

I know someone who thought that instead of opening mail from the Tax Department, he should write 'deceased' across the front of the envelope and return it unopened to the Tax Department. The first time he didn't spell 'deceased' correctly and they tried to quarantine him. The second time it backfired as they contacted his next of kin, his elderly mother, and upset her somewhat when they asked just when her son had been killed.

Become a subcontractor.

Use of a 'contractor's contract' may offer advantages in breaking away from the PAYE treadmill. Advantages are:

- Control over your own money and ability to earn interest on it until assessed.

- Benefits of tax deductibility on many items, including car, phone and portion of your home as office.

A disadvantage may be in paying provisional tax, but only if you pay it on time. The success of the use of such forms depends on your particular circumstances and the number of others you can encourage to simultaneously participate (advice should be obtained).

There are legal tests to define 'contractor' or 'subcontractor' and one Tax Department decided to question the validity of prostitutes using these 'contractor's contract' forms and breaking the Tax Department's automatic access to the girls' money collected by the madam. The case went to court where the judge agreed with the girls' defence that they, in fact, complied with the legal description of subcontractors:

i.   Could they complete the task without constant supervision from their employer? – YES

ii.   Did they have control over the standard of workmanship? – YES

iii. Did they bring their own tools to the job? – YES

Always encourage any rumours that indicate your ownership of any assets not really owned by you. The fun starts when the taxman tries to seize these assets.

In their job, as tax investigators (or 'fiscal fiends', as well-known tax rebel Peter Clyne calls them), they must feel pretty rotten about the way they are treating their fellow Australians. It may help them, from time to time, if you enter into stimulating correspondence, along these lines:

Dear Mr Deputy Commissioner,

I am acknowledging the unsolicited literature that you sent to me, dated 16th March 1976. Among other things, you allege that I 'owe' a liability and request that I 'pay' it; you claim that I have the right to appeal against your finding after I 'pay' the alleged liability.

The mentality that takes such positions as these used to puzzle me. However, it does no longer, because I have long since concluded that and, my dear Mr Deputy Commissioner, I am sure you will agree with me, we live in a highly irrational society. In an irrational society, the incompetent and dishonest backstab themselves into positions of responsibility and prestige, while the competent and honest usually make it on their own merits. Which means, in an irrational society, people in high places can be expected to do irrational things. Irrationality, Mr Deputy Commissioner, is a euphemism for stupidity. Let me use you, as an example, to explain that. You make several requests that I 'pay' the tax, 'pay' the amount, make payment, and so forth. The words 'pay' and 'payment' imply a transaction in which something has been exchanged and expected to be paid for. In other words, when you request that I 'pay' a certain amount, you imply that you have performed some service for me or given something to me. I am at quite a loss as to what it is that I should pay for. To the best of my knowledge you, the Federal Government, has or does nothing that I need or want or that I haven't already overpaid for in fees or indirect taxes.

If you maintain that I 'owe a liability' to you, I request that you demonstrate exactly what is in my possession, what it was that I requested, or what I used that was provided by you, during the periods of time in question.

I will grant that the government does things for me that I neither requested nor need. However, since I did not request them, you must regard them as acts of charity and not expect me to pay for them. I won't pay for them, for the simple reason of self-preservation, otherwise what is to stop every crackpot in the country from packaging a load of garbage, bolting it to my car, throwing it in my window, or hanging it about my neck and then sending me a bill for services rendered?

I am sure you want to consider them as acts of charity. Consider what you would concede if you don't.

If you refuse to identify what it is that you provided me but still allege a liability on my part, aren't you trying to affect a rip-off?

If you identify something that is not, nor ever was in my possession, then how can I possibly 'owe' you anything? If you still maintain that I 'owe' a liability aren't you seeking to make me pay for something someone else consumed?

That would make you the agent of a parasite.

If you can identify something that you gave me, during the periods of time in question, then I will give it back and that will cancel my alleged liability. If you subsequently decline to regard my alleged 'liability' as cancelled, then you admit either that you did not give me what you claim to have, or you did give me something but it is of no value. The first admission automatically cancels my liability, since nothing was exchanged and the second admission makes you a hawker of shoddy goods.

Also, my dear Mr Deputy Commissioner, that puts you in a rather sticky situation. It makes you either a crackpot, a rip-off artist, an agent of a parasite, or a hawker of shoddy goods.

The literature you sent me states that I have a right of appeal from your finding. This implies that a judgement has been rendered against me; which implies that I have been joined in a trial by jury and that I have presented my case therein.

Alas, however, this is not so. I have presented my case nowhere, no jury has rendered a verdict and no judgement has been rendered. Your finding does not represent due process of law, it is a mere opinion, legally void, unenforceable and affords no protection for acts done under it. Since your finding does not constitute a legal judgement it is legally non-existent. There is no court in the land that has jurisdiction of an appeal from a non-existent judgement.

This alternative that you offer me is clearly impossible – apart from the rejection of it. Anyone who claims that I may appeal against non-existent judgement has certainly lost touch with reality or is incapable of distinguishing between one's feelings and reality and this is a defining characteristic of psychosis.

I strongly suggest that you re-examine your position.

You allege that I owe a liability. I do not. You imply that I must comply with your rules which are legally void in order to prove my innocence. I will not.

You request that I pay the alleged liability or appeal your findings. I will do neither. I shall completely disregard your requests and with your co-operation present my case before a jury of twelve.

You remain, Sir, my humble obedient servant.

H.J. Van Groenwoegel

Reluctant Taxpayer'

While the Politicians/Bureaucrats Are Busy Shuffling Deck Chairs on the *Titanic*, What Can We Do?

Things that won't work

Someone asked me the other day "yes, but aren't you going to have absolute chaos if people start refusing to pay taxes?" I replied, "Even with all this tax money supporting governments, dedicated to total central planning of your life and mine, haven't we got chaos now?" If blindly paying taxes like 'good little citizens' has bought us chaos, it is now important that we do something about it, even if only on a personal, individual level.

First, I would like to suggest ways that won't help. They might look attractive, but history has shown that they just won't work, even though the people who have tried and are trying them now may be well intentioned.

What they are trying to do is make socialism work 'Liberal Party style' or make it more efficient. By suggesting that the 'mixed economy' is a valid alternative, all that will be achieved is a continued drift to socialism.

Many of the world's top economists at the Mont Pelerin Society Conference in Hong Kong in September 1978 made it very clear that if you aim for a 'mixed economy', i.e., a mixture of socialism and free enterprise, you will end up with socialism. However, if you aim for a pure free enterprise society you will still end up with a degree of socialism but it won't be serious enough to bring the country to a grinding halt.

Many of Australia's politicians appear confused and treat politics as if it were economics. Economics is the science dealing with the production and distribution of wealth, subject to natural law, while politics is the art of restraint, subject to expediencies. As long as the present fashion of treating economics as a branch of politics continues, the current confusion among most politicians and some economists will also continue. This confusion among politicians has contributed to their declining prestige on the social scale, to the extent that a well-known mineral economist told me, recently, that he has devised a foolproof way of telling when a politician is lying: He just watches to see if their lips are moving!

Sometimes one is forced to doubt whether it is still honest stupidity, rather than sinister intention, that leads politicians to invert the truth …

I prefer to believe that it is doctrinaire blindness rather than a devious attempt to destroy the existing order which can make a politician deny the obvious truth.

If this doctrinaire blindness leads to incompetence, then I suggest that this is the most dangerous form of incompetence in existence. If I'm incompetent in writing a book, that's okay, people will refrain from buying and no-one is hurt but me. On the other hand, if politicians and bureaucrats are incompetent, we can't voluntarily sidestep the consequences of their incompetence.

The least we can do is withdraw out financial and physical support. Let's not spend any time or energy on trying to make socialism work, as Leonard Read so aptly puts it, "The only thing which gives socialism the appearance of working is the freedom that socialism has not yet destroyed."

Things that might work (Tax Reform)

Consumption Tax

One of the basic iniquities of our present system, with its graduated, escalating tax scale, is that it is based on how much a person contributes to the economy. He should get a prize for contributing, not a penalty. This is why our present system is destructive and is nothing more than punishment for pure productive ability. This could even be dressed up and sold to the electorate by our politicians as being a way of tapping the high-living jet-set for more tax and giving a break to humble folk trying to squirrel away a portion of their hard-earned wages.

There appears to be a mental fixation in many minds that the graduated tax scale is fair because they think it is based on 'ability to pay'.

That phrase always sounds good to people who think of themselves as lower or middle-income earners. It makes them feel good that someone who is earning more than they are is really paying more tax than them. In practice it doesn't work that way, but the politicians like to encourage such myths.

It is not good trying to argue against the concept of 'ability to pay' by using long scholarly economic arguments which, while correct, demand more time than anyone has these days. It may be better to use the phrase 'ability to work' and ask if taxing by 'ability to pay' is cheating those with the ability to work.

This might help focus attention on the more just concept of a 'consumption tax' – that is, where people are taxed on what they consume, generally along the lines of pay as you use, for all services provided now by governments and misleadingly called 'free'.

Encourage and expand voluntary organisations

Whilst the concept of the government confiscating our earnings by force is dubious morally and uncomfortable economically, there remains the problem of those in genuine need. It is unfortunate that the long-established and highly efficient voluntary charity organisations in Australia have been forced to become something of a poor relation to our government's monolithic Social Security Department.

There is merit in drawing attention to the essential motivational differences between the government welfare agencies (compulsory) and private charity organisations (voluntary). Highlighting the different management goals and styles between compulsory and voluntary organisations and their respective abilities to 'solve problems' might help the latter retain their independence.

It should be remembered that if we are at all successful in our efforts at Tax Resistance, then each of us has the right to do good at our own expense.

In a sense, the success of the whole tax-revolt individual-responsibility movement may depend on the willingness of the citizenry to resume private responsibility, voluntarily for genuine cases; such as those who have become wards of the State, whether they be the elderly, the poor or the disabled.

Curing Government Bloat

Unless we prevent politicians from concealing costs from those who pay them, we will almost certainly continue to have government of the people, by the politicians specifically for the interest groups.

Government grows so fast because the costs of government are diffused among the taxpayers at large, while the benefits are often concentrated among a relatively small number of people. We can call them 'special interest groups'.

The beneficiaries have more to gain from the expansion of government programs so they 'shout the loudest'. Politicians hearing this 'shouting' feel they have more to gain from proposing or supporting increases in these benefits to the few than from representing taxpayers.

There is no quick or complete cure for this condition and it is an area which needs considerable research and education. The study of what economists call 'Public Choice Theory' explains clearly that the 'concentrated benefits to the few' always emboldens those 'few' to lobby the hardest and to disburse the costs over 'the many'.

Some areas of action are:

i. Identifying the manner in which the government uses inflation for its own benefit, by the manner in which inflation forces taxpayers into higher and higher tax brackets with government being the only beneficiary.

ii. While company tax is a major source of tax income, to some extent this tax is treated only as a cost of production and passed along so that it is eventually paid by the consumer. Since it is concealed in the price of goods and services we buy, we don't realise that we are paying it.

This point should be brought out as often as possible to explain the argument for gradually substituting other sources of revenue in place of company tax.

There is a growing school of thought in the U.S.A. that company tax should be quickly abolished and that any additional company income, after price reductions, should be allocated to individual shareholders who would pay tax on this income at their own personal tax rates.

The other side-benefit would be to alleviate difficulties companies are experiencing in raising equity capital.

The yield on most company shares is not that great, partly because company tax reduces the amount of income available for dividends. If company tax were abolished, companies would pay out much higher dividends, thereby making stock ownership more attractive and easing the capital formation crisis.

As all the government's experiments in taxing us and regulating us fail, it will become clear that the only lasting way to increase the standard of living of people is to increase the amount of capital invested per capita. Abolition of company tax would encourage such investment.

iii.   Federal grants to State and Local Governments constitute what is, in effect, a device for getting around the resistance of State and Local voters to higher taxes. Taxes ae more effectively lodged at the local level where the taxers have to personally confront the taxed. The taxers are then more prudent about spending the money as the taxpayers are often watching more carefully.

iv.   A movement to be started toward 'truth in government', making it obligatory that any level of government may not enact any measure entailing public expenditures unless it is accompanied by a rigorous analysis of costs projected, say, five years into the future.

Flat Tax Rate

Escalating rates or graduated tax scales increase the awareness of the burden of taxation and consequently, lose revenue for governments through loss of incentive to work and through tax avoidance. There is, therefore, a strong case from both the government's point of view and the taxpayer, for the shift to a Flat Rate Tax. This should not be confused with the present government's three-tier escalating rate.

It may not be a coincidence that two of the countries with the lowest inflation and highest productivity, Switzerland and Hong Kong, both have a flat rate tax of approximately 15 per cent. I understand that this rate of tax is only levied once, i.e., if tax is paid at company level, the same money is not taxed again when distributed to individuals. Many people overseas are quite horrified to hear about our Australian system of taxing the same money twice. In Switzerland or Hong Kong there is virtually no such thing as a tax accountant. Who wants to cheat on tax when it is only 15 per cent?

In Hong Kong no tax is levied on incomes below about $8,000. This overcomes any objection about a switch to a Flat Rate Tax causing low-income earners or pensioners to pay more. In Hong Kong, the population of 4.5 million only has about 142,000 actual taxpayers, so the cost-effectiveness of the taxes the government does collect is impressive. There exists much, well-researched material available to substantiate the validity of the benefits of a Flat Rate Tax. Another model country for the Flat Rate Tax is Estonia.

Constitutional Limitation on Government Spending

Whilst it is desirable for the people to limit their government's budget, a fundamental problem is that we have no means whereby the public ever gets to vote on the total budget of the government. There always appears to be an infinite number of good and desirable proposals, but we lack the device to limit government expenditure, often being encouraged by the very same people who are the first to criticise governments.

A solution was suggested by the Professor of Economics at Virginia Polytechnic, David Friedman, during a visit to Australia in 1978. He said that government spending should be a fixed percentage of Australia's income. There should be a referendum, at every election, with voters giving their preference for a 10 per cent reduction or increase in government spending. He said the move would not be popular with politicians, who equate spending with power. However, if it came to a choice between small government or being in opposition, they would choose a restricted spending program.

The Things That Work

We need tax relief, not tax reform.

> When a government becomes powerful, it becomes destructive, extravagant and violent. It is a usurer which takes bread from innocent mouths and deprives honourable men of their substance for votes with which to perpetuate itself.
>
> -- Cicero, 54BC

The politician's definition of tax reform is, simply, changing loopholes into nooses. While there is much talk about reform, there is certainly not much benefit to flow to taxpayers.

The Only Way

What Australia needs is tax relief, not tax reform. The only way to get this and the only way to get better value for your tax dollar is to reduce the size and power of our government.

I see four broad areas of participation where constructive efforts can be made to limit and reduce the power and size of government. Individuals of different talents and temperaments would be drawn to one or perhaps more of these activities as a way of participation.

The four areas are, in summary:

Education

The current level of public awareness in Australia is one real cause of our problems and reminds me of a salty section of land on my Esperance farm. Nothing would grow on it until I imported some salt tolerant seeds from Turkey. This *puccinellia*, as it is known, after a few years growing on the salty desolate patches, has rehabilitated this land to the stage where productive palatable pasture has germinated.

The government's mass-produced education system only teaches us what the government wants us to know. So, at the moment, the intellectual vacuum existing among so many people makes it difficult for the seeds of better ideas to take root.

If you have better ideas, the educational process is one way to cultivate the ground and then, if the seeds don't take, you know that you're 'planting techniques' need revising.

An encouraging sign is the increased availability and quality of educational self-improvement material. This allows any of us to increase our ability to recognise the economic and moral errors of government intervention in our lives and the economy.

Deregulation and Freedom of Choice

This calls for active participation and logically follows from the educational process. Individuals and organisations can focus attention on the high costs to consumers of existing government regulations.

The point which should never be missed is that regulations rarely benefit the consumers. Regulations benefit the regulators by expanding their power base and they benefit the regulated industries by restricting competition.

The moral implications of regulations are, once again, as serious as the economic implications.

Any law that favours one person or group at the expense of another is not a just law. If this is correct, it casts doubt on much of the special interest legislation continually being produced by parliaments in their efforts to appease special interest groups. We should all be treated equally under the law and our existing law courts should be adequate to deal with criminal damages or fraud actions.

The concept of freedom of choice for transport, airlines, mail services, etc., are among the basic freedoms that have been slowly eroded. There is so much logic to substantiate the promotion of freedom and competition that they are ideal topics for private or public debate.

A classic example is our lack of freedom in choice of currencies. It is bad enough that our government is debauching our currency through inflation. Far worse is that they have made it illegal for us to protect ourselves by choosing to hold more stable currencies.

The extent to which people will go to avoid holding Australian dollars is shown by the latest trick. [Note not valid at 2019 but an interesting commentary on the 1970s]:

- Purchase an overseas air ticket.

- This permits the ticket holder to buy up to $4,000 in stable currency traveller's cheques (Swiss Franc, D'Mark or Yen).

- Cancel their air ticket but keep the traveller's cheques, regarding them as safe as gold but more convenient.

Freedom of choice of currency may be the only power in the hands of the people that will force governments to issue and maintain an honest currency. It is interesting to note that inflation is minimised in countries where currency freedom is maximised. Two examples are Switzerland and Hong Kong.

Tax Revolt and Tax Resistance

Tax revolt is best done along the lines of constitutional limitation on most governments' unlimited powers to tax and confiscate.

Governments will be more effective, in the long term, only if equal billing is given to the moral justification for tax resistance as well as the more obvious economic reasons.

In Australia the level of intellectual debate has not yet matched that of the U.S.A., despite introductory seeds being planted, in particular by *The Australian* and the *Bulletin* on the east coast and the *Sunday Independent* on the west coast.

In the U.S.A. many heavyweight economists and intellectuals have given logic and credibility to the Tax Revolt, highlighting the fact that the tax limitation laws being enacted are only stop-gap measures which are holding back the tide until public opinion moves in the direction of limiting the powers of government to the role of umpire or referee rather than Big Brother!

There is ample evidence of a definite movement of public opinion toward greater scepticism about government programs. People are aware that they are not getting their money's worth through government spending. They conclude that many government programs have not had the results intended by their supporters. However, it takes time for such ideas to be accepted by the politicians who, after all, are mostly elected followers and not leaders of public opinion.

Professor Milton Friedman directs taxpayers' eyes to what government spends and states: "I believe, along with Parkinson, that government will spend whatever the tax system will raise, plus a good deal more. I am in favour of cutting taxes under any circumstances, for whatever excuse, for whatever reason."

Dr Arthur B. Laffer, a former Treasury Department economist and currently editorial advisor to the *Wall Street Journal*, has formulated a graphical representation that is helping shape the opinions of taxpayers and politicians alike.

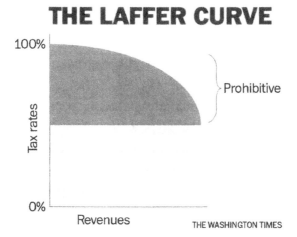

In June 2019, Arthur B. Laffer was awarded the Presidential Medal of Freedom for popularising this concept for demonstrating that more tax is often collected at lower tax rates.

Adelaide's Prof. Jonathan Pincus (then working in the U.S.A.) worked with Arthur Laffer in perfecting these studies.

The 'Laffer Curve' illustrates the old law of diminishing returns by showing how the government can collect the same amount of tax by two methods, i.e. zero tax collected at both 0 per cent tax rate and 100 per cent tax rate.

Tax Revolt on a public level is still viewed, by many, as a hopelessly exhausting occupation. True, much public protest bears no fruit for years but sometimes, many years later, its productivity can explode. Take, for example, the writer Aleksandr I. Solzhenitsyn. His aim was to expose oppression and injustice under communism in the Soviet Union. At that time his public protest had value in its potential to change people's minds with a new idea. It had considerable educational impact.

His book *The Gulag Archipelago* was first published in 1973.

Today, nearly 50 years later, his book is more relevant than ever before because of the widespread teaching of socialism and Marxist philosophy in western universities. It has now been republished, combining all three volumes into one.

Political Parties

There is an encouraging increase in scope for working within free enterprise, limited government, limited tax political parties, such as are now growing rapidly in many countries of the world, i.e., the Libertarian Party in the U.S.A. and the Progress Party in Australia and Denmark.

The first International Conference of Libertarian Parties was held in 1978 with representation from 17 countries. Their support comes from all sectors of the community as they provide a real alternative to the other look-alike, act-alike parties. These limited tax parties take a firm stand on the dubious nature of compulsory taxation and believe that the individual, as the primary agent of economic transactions, ought to be free to spend what he earns, as he prefers.

They advocate a voluntary society in which the proper function of government would be only the protection of life, liberty and property of each from force, fraud or coercion. Most remaining government services, therefore, would be on a fee-for-service basis with little or no spill-over into 'consolidated revenue', thus limiting government vote-buying activities. Government services would have to compete in the marketplace for the consumer's dollar and government activity would occupy only that share of the market which reflected the choice of consumers.

Once again, it takes time for the repeated failures of government intervention to influence the growth in public support for these new limited government parties. However, in the short term, if such limited government policies are effectively and persuasively expressed in words, they may be 'stolen' and used by the major parties. In that way at least a few more seeds of freedom may be propagated.

Summary

Unfortunately, most of our problems would not have occurred had we been as concerned about preserving liberty as we were about earning a living and paying taxes. Most of us are so busy and preoccupied that we have neglected to preserve the freedom that protects our rights to keep what we have earned.

To conclude, let me state the obvious. Things will not remain the same. They either get worse or they get better. If Australians are not successful in reducing the size and power of government (remember the government will not voluntarily shrink), then I see a fairly grim scenario developing.

Through the adjustment of their individual activities, Australian citizens will attempt to escape from coercion, by changing their private behaviour, rather than by attempting to influence politicians. Such adjustment, itself, will have an indirect political impact. Suppose, for example, that coerced citizens respond to penal taxation either by increasing their consumption of leisure or by increasing their tax resistance, thereby reducing the tax base available to the government. If tax revenues are to be maintained, government must then levy yet higher tax rates and thus impose additional coercion on many other citizens, thereby generating further political participation in opposition to its policies.

If the government doesn't shrink, as a result of its reduced tax base, it will continue to raise revenue elsewhere by a greatly expanded network of inescapable, indirect taxes, e.g., a Goods and Services Tax (GST), a Value Added Tax (VAT) or a Retail Tax on Goods and Services.

Once again, the result will be the opposite to their intent and instead of raising more revenue there will be a rapid escalation of barter and smuggling in the form of a most sophisticated subterranean economy. Many other countries have this now and it is great news for the accountants as most businesses operate three or four different sets of books – one for each special occasion.

Many people will not wish to identify themselves with the continued drift toward being a country that is a paradise for potential malingerers and hell for people who wish to work (without slipping onto the subterranean economy) and they will exercise their democratic right to 'vote with their feet' and leave Australia.

In 1977, for an article titled "The Alienated Australians", I interviewed 27 Australian businessmen now living overseas. They left Australia "as a result of the rapid increase in the heavy-handed bureaucracy". Rather than allow themselves to be kicked around the groin, on a daily basis, they have elected to live elsewhere in an environment of greater personal and economic freedom. Others will recognise the benefits of 'internationalising' themselves. There is no country that is simultaneously the best country in which to live, to invest and to work. The solution is to select a different country for each of the three functions of living, investing and working.

On the other hand

For those of us who seek an improvement in the 'human condition' there is a tremendous challenge to pursue our own chosen plan of action.

The fact that we are only individuals should not deter us. Solutions won't come from the masses, for the simple reason that the individual is, and always will be, more important than the mass, just as the soloist is more important than the chorus. There may be fine voices in the chorus, worthy of being soloists themselves, but they are lost in the mass. The mass, is merely an agglomeration of individuals, without the possibility of being heard individually. Whereas the solitary person is capable of thought and self-expression. That is why libertarians often feel 'lonely'!

No idea, good or bad, has ever emanated from the minds of a multitude. Ideas come to one mind at a time. People, it has been well said, go mad in herds, while they only recover their senses slowly and one by one. The challenge to us, as individuals, is vital for our very wellbeing. If economic sanity can be restored, we can look forward to the benefits and pride that will come from living in a great country with unmatched potential for enterprise and individual initiative. By working toward a reduction in the size and power of governments we are maximising the freedom of individuals and their creative capacity.

Only by maximising our freedom to produce, trade and exchange will we reach our true potential as individuals working for our own individual goals. Sometimes personal profit, sometimes love of a challenge, sometimes love of our country and the freedom one could enjoy in it, sometimes the satisfaction of seeing human beings released from poverty.

Those who profess to favour freedom and yet deprecate agitation,
are men who want crops without ploughing the ground.
They want rain, without thunder and lightning.
They want the ocean, without the awful roar of its waters.
This struggle may be a moral one; or it may be a physical one;
but it must be a struggle.
Power concedes nothing, without demand.
It never did and it never will.
Find out just what people will submit to and you have found out
the exact amount of injustice and wrong which will be imposed upon them;
and these will continue until they are resisted with either words or blows, or
with both.
**The limits of tyrants are prescribed by the endurance of those whom
they oppress.**
(Negroes will be hunted in the North,
and flogged in the South,
so long as they submit to those devilish outrages,
and make no resistance,
either moral or physical.)

– Frederick Douglass; 1818 – 1895
Black American slave who became an orator, writer & statesman.

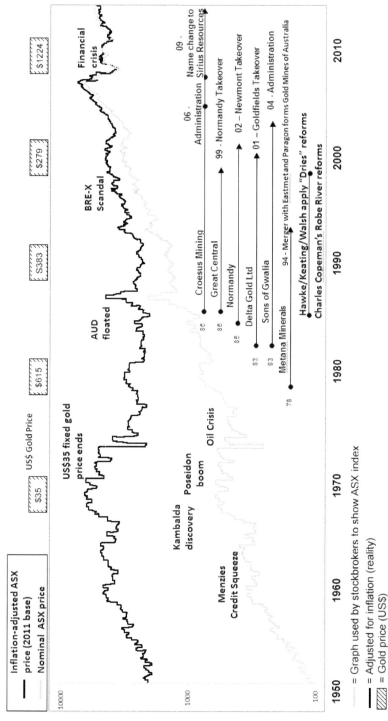

"The history of the past 50 years is like a persistent shadow, it just keeps following us around."

Ron Manners, Geoffrey Blainey AC, Sir Arvi
Parbo AC and Ross Fardon enjoying lunch.

Roy Woodall stands on the exposed
nickel discovery gossan

The Kambalda Commemorative Reunion
Group – September 19 1992

8

TRIUMPHS AND TRAGEDIES OF AUSTRALIA'S EXPLORATION AND MINING INDUSTRY:
1960s — 2015

Heroic Misadventures, my last book, was subtitled *Australia: Four Decades—Full Circle* as I was concerned then (in 2009) that Australia had emerged from an extraordinary mining boom, wallowing in debt — *Heroic Misadventures* proceeded to build the case for my concerns which may have not been obvious then but with the passing of years and the advantage of hindsight are now obvious to all.

So when I received an invitation to speak alongside a group of interesting, well-informed 'participatory observers', for a 'retrospective analysis' (30–31 March 2015) on the significance of lessons learned over the past 50 years and, more particularly, of how to make these experiences useful for the future, I enthusiastically accepted the invitation.

So, although this is essentially a personal outpouring of how the dramatic effects of the last 50 years affected my life, it is also an overview of Australia's exploration and mining industry over this same period.

Our lives over this period have been inseparable from the fortunes of our industry and the ways in which it has taken shape, blossomed, and let's face it, found itself dashed upon the rocks, shipwrecked and friendless from time to time.

To many of us, this history of the past 50 years is like a persistent shadow, it just keeps following us around.

The story that I'll tell you *is* a tale of triumph and tragedy and I'll finish with this question: what are we going to do with all our accumulated wisdom as we look forward from 2015?

If you don't see any reason for me to use a word such as 'tragedy' to describe our industry, I will argue that there is no other word that expresses my feelings as I have watched our industry lose its comparative advantage over this 50-year period. It is a double tragedy because we have done this to ourselves, often with the best of intentions.

So, let me give you a road map of where we'll go in this probing adventure.

Road map

- How Australia was as boring as 'batshit' in the 1950s, leading up to being 'king-hit' by the Menzies Government's credit squeeze of 1961–62.

- The nickel discovery, which came upon us slowly at first, then gathered pace as we rode the wave!

- My personal story as 'captain' of several small companies as I piloted them through very stormy seas.

- A question: Where were our leaders and philosophers

when we had a chance to set some sensible rules on how the game should be played? How we lost our comparative advantage.

- Our own useful investment in the future. What are we going to do with all our wisdom and data?

The boring 1950s

Just in case any of you don't realise the significance of what you were doing back there in the 1950s/60s, I'll tell you how boring my own hometown Kalgoorlie was before explorationists kicked a few goals.

Following secondary school I tried to sign up for geology at the Kalgoorlie School of Mines, but the director of the school R.A. (Hobby) Hobson said: "Ron, mining has finished, so it would be far better for you to do something more involved with engineering".

I had always been passionate about anything electrical, so I became an electrical engineer.

I sat through several years of those 7.00 pm – 10.00 pm lectures, five nights per week, with only one finishing hour per week at the Palace Hotel.

My classroom colleagues, battling away in similar fashion, were an interesting bunch of rugged individuals. Among them were Keith Parry and John Oliver, both of whom played a major part in my life journey.

There was a wide range of age groupings at the school then, as all of us had day jobs.

My day job was in our family mining engineering business.[15]

---

15 We celebrated our 120th anniversary of doing business in 2015 and I must say right now that I'm enjoying the third time in my life when I feel that I have got it just about right with team spirit and performance.

The electrical engineering choice turned out to be fortuitous, as our family company was struggling (just as the whole region was struggling under a government-imposed fixed price of gold, $35 per ounce, for many years). We survived by converting the Goldfield region's steam-driven mine winders to electric, featuring very sophisticated electrical systems of over-speed and overwind protection, a major improvement from the earlier steam-driven winders.

However, this came to an end, like so many jobs today, doomed to extinction and we moved into a demoralising period when shafts were being closed, almost on a monthly basis, and in many cases the winders and headframes were being disassembled and shipped off overseas. I recall that some went to Brazil.

Our business was barely viable, so my father and I bought a small farm at Esperance where we took turns to run sheep for a couple of days a week and then serviced the Norseman mining area through a small branch in the main street while still managing our business in Kalgoorlie, backed up by excellent staff.

Most of my school friends had managed to escape from Kalgoorlie and head for greener pastures. In the hope that Australian mining would see a resurgence of gold mining, some of the remnants still left in Kalgoorlie attended the occasional prospector lectures. I remember one by Sam Cash on "Loaming for Gold". These notes I referred to several years later.

Gold mining or prospecting was hard to imagine as something that would ever provide one with a livelihood, but the historic romance of it all maintained our interest.

So, that was the backdrop when disaster actually struck, in the shape of the Menzies Government's Credit Squeeze in 1961–62, when by one simple edict they instructed the banks to reduce all business overdraft limits by 50 per cent.

The Government's motivation was that they saw signs of prosperity, so they needed to rein it in. Perhaps this was about the time that I started developing a strong sense of cynicism towards all government actions and began suspecting that the end results of their activities were often the opposite of their stated intentions.

The Credit Squeeze brought on the demise of most of Australia's major retailers. National chains of stores such as Cox Brothers, Reid Murray, H.G. Palmer and Eric Andersons simply closed their doors because there were insufficient hire purchase facilities to keep going.

Most of my memories now from the '50s, the happy ones anyway, were simply about discovering jazz, discovering girls and thinking how difficult it was to be intelligently creative in business when for so much effort you actually had so little to show for it.

Well, with all that boredom, along with a government-fixed gold price of US$35, what else could go wrong? The whole region was desperately in need of a visit from the proverbial 'tooth fairy'.

The new nickel era started slowly

The 'tooth fairy' arrived in the form of an exploration revolution. The revolution was started by Roy Woodall and his merry team at Western Mining Corporation. There were stirrings around the Goldfields that something was about to happen.

I remember, in late 1965, that an important figure in the state business community (it may have been the Chairman of Wesfarmers) wrote to my father and said, "Charlie, I'm hearing good things about Western Mining Corp, what sort of a company are they?"

I was with my father when he was reading the letter and recall him giving me some advice on what to do when people ask for opinions in this way.

His first comment was gather the facts. So, we both walked across the road to Ron Reed's office (Kalgoorlie stockbrokers) and obtained a copy of Western Mining's latest report and I recall my father replying along these lines: 'From the balance sheet they look as though they are going broke, but the people are sound and they know what they are doing'.

A few months later, Dad and I were in Perth at a function and this same correspondent marched up to Dad and said: "Thanks Charlie for your advice, that investment has turned out well for me."

Dad replied: "I thought I may have put you off with my concern about their balance sheet?"

I can clearly remember the gentleman replying: "From my experience, in Western Australia, balance sheets don't count for much – it's all about the people". I often think about that comment whenever I'm doing business with anyone and it has worked out well for me.

Two of the three nickel booms
(the third occurred 17 years later)

There were two distinct phases to Western Australia's appropriately named nickel boom.

Some people object to the name 'nickel boom' but it totally transformed the sleepy back blocks of Western Australia, with overflow effects on the entire nation. The world took notice and started to arrive in plane loads. Anyone who could string a few words together became an interviewable celebrity.

Kalgoorlie went from having a few remnant geologists to having the highest number of geologists of any city in the world, with New York City running second. 350 exploration companies beat a path to our Goldfields area motivated by the fear of missing out. How many of those 350 actually made a profit? Two ... Western Mining

Corporation and Metals Exploration at Nepean.

I say there were two distinct phases to the boom, the first being the Western Mining, Kambalda discovery.

This discovery did not amaze people as it was seen to be done by a group of highly professional technologists who had been beavering away for many years and had accumulated patience and skills. They had confidence in themselves and the community had confidence in them. That community confidence probably held the region together during all those difficult years.

Then, there was the Poseidon nickel boom three years later. This discovery appeared to be made by a two-cent company run by Norm Shierlaw, a little-known Adelaide stockbroker. They had seemingly driven a drill rig a few kilometres out from Laverton, dropped a few holes and discovered what was thought to be another Kambalda world-class nickel region.

This looked to have been achieved without much technical input and that's when the proverbial shit hit the fan.

All those companies now based around Kalgoorlie, that had been searching for nickel came under intense pressure from management and boards. The refrain was "if this little two-bit company can do it, then so must we". Exploration budgets were doubled and the media attention immediately quadrupled.

How I lived through those next few years I'll never know.

Prospecting days

Suddenly my prospecting skills went from hobby status to being extremely useful as I set out to peg half the state, optimistically thinking that this would go on forever.

Prospecting became an all-consuming passion for me, and I was fortunate in developing a wide selection of partners with whom

I enjoyed working. They included some aboriginal contacts made from my earlier primary school days in Kalgoorlie. I learned of their skills in seeing subtleties in the rocks with their naked eye which our geologists appeared only to see with the assistance of magnifying glasses or microscopes. So, I equipped several of them with old wooden cigar boxes, suitably partitioned, containing a range of nickel gossan samples to give them an idea of what we were seeking in our search for nickel prospects. This resulted in a degree of success and if, after a first-pass inspection of the ground there was some significance in their gossan discovery, I paid them a flat fee of $10,000.

At that point of minor encouragement, I brought in one of the local geologists. I would hire people such as George Compton to do some of the reconnaissance work. If the interest was maintained, the ground was then pegged. If there was insufficient merit, then it wasn't pegged. The aboriginal prospectors received their $10,000 and I took my chances. Such is the nature of entrepreneurship.

My favourite prospecting partners were John Henry Elkington and his wife, Norma. John had commenced geological studies but diverted himself into prospecting as a way of leapfrogging, as it were, his geologist brother, Dick Elkington. He desperately wanted to find a mine before his brother did.

John and I acquired many significant areas over the years, and this led to a series of corporate adventures. Our prospecting company was aptly named Mannelksploration Pty Ltd.

We had developed an affinity for the Forrestania area and were actively involved in securing some strategic ground there when the State Government announced a pegging ban. Can you imagine that? In the middle of a full-scale nickel boom their registration bureaucracy fell behind in their record-keeping, so instead of speeding up their own process and solving their deficiency they instead announced a pegging ban.

Well, that brought to a halt the momentum that had been building up and all we could do was continue prospecting and 'sit on' the results until such time as the ban was repealed.

The State Government gave notice that on a certain day they would be making an announcement over the ABC shortwave band that the pegging ban had come to an end. There were whole battalions of prospectors and geologists all stationed out on their special prospects on which they had been working, during the time of the ban, waiting for the 'blips' to come over the radio and at that moment it was on for young and old. Fortunately, Mannelksploration acquired some prime ground at Forrestania which is still being actively explored/mined. Preliminary work was done there by our joint venture partners, AMAX Inc, the major U.S. mining company.

We successfully prospected the Forrestania region for some years, resulting in pegging the Maggie Hays Hill (later the basis of the successful flotation of the public company Theseus Exploration NL. Incidentally, history tells us that Theseus was the illegitimate son of Poseidon). From Maggie Hays Hill we then continued working down the Bremer Range formation from the north, finding this geological environment fascinating. (Everything else was covered with sand.)

I was reminded about a certain incident, at well-known geologist Ed Eshuys 70th birthday party, when Ed related the following story, that stirred my memory.

Ed, at the time, was the manager for the Belgian Government's exploration arm, Union Minière and they were also wildly enthusiastic about the Bremer Range region and frantically prospecting and pegging up from the south. They had a substantial team, far out stripping our humble 'do-it-yourself' team and they were, with great speed, digging the prescribed trenches and hammering in the required pegs (of those days).

They were certainly working with great speed but, being typical government employees, right on the dot of 4.00 pm they promptly packed up and headed back to camp.

You can imagine the sheer delight and ecstasy that we experienced when working down from the north we came across all these ready-dug trenches with the tenements all marked out and pegs all in place but, of course, no papers were fixed.

Being of an extremely tidy temperament we felt that no job should remain incomplete while the sun was still shining, so we continued merrily papering all these delightfully positioned pegs. Our enthusiasm was so great, in fact, that we worked well past midnight, by kerosene lantern, to ensure that our task was completed. Actually, there was added job satisfaction for us as we lurked in the bushes and witnessed the absolute cries of anguish from the Union Minière team when they arrived the next morning.

However, it wasn't long before we, Mannelksploration Pty Ltd, received a summons to appear in the Norseman Warden's Court, accused of having performed something of a dastardly act. John Elkington and I had Kalgoorlie solicitor, Tom Hartrey – the well-known political orator – defending us on this occasion and, as usual, his courtroom performance was spellbinding.

There was always excitement when crowds gathered to witness Tom's eloquence at its best. He proceeded to assure the magistrate that this situation ranked far above the humble Warden's Court and in fact was a matter of international natural justice. Union Minière had been downright careless; he likened them to a careless fruit tree owner who deliberately grew ripe fruit over his boundary for the benefit of all those passing by.

Neither the Warden nor Union Minière's attending solicitor were up to this full-frontal confrontation by Tom Hartrey and we walked away with the prize. Naturally we celebrated well into the night and this secured ground became the basis for the float of Kalmin Exploration Ltd which we did not realise at the time was the last company floated out of the nickel boom. The hysterical aspects of the boom really only lasted for about 18 months then collapsed with heavy breathing and gurgling shortly thereafter.

Kalmin Exploration Ltd then transformed itself into Bellcrest Holdings Ltd which became one of W.A.'s active home builders but, more importantly, Ed Eshuys has forgiven me for this 'pegging mischief' and we remain firm friends to this day.

The first ground I ever pegged was at Siberia (north of Kalgoorlie) and it went into a new float called Westralian Nickel NL. They soon appeared to have found another Poseidon/Windarra, causing their shares to rocket to $8.80 and I copped a big fat tax bill based on that figure (accurate at 30 June of that particular year). However, my shares were escrowed and unsalable. I ended up selling them at 15 cents each the following year.

The Australian Taxation Office (ATO) showed little interest in my predicament, so realising that they had no comprehension of natural justice, I simply had to leave the country as they were repeatedly sending me interest bills on the large outstanding amount.

I kept busy overseas with several jobs, such as running a hotel in Bali and working for an up-and-coming merchant bank called Nugan Hand.

Nugan Hand's specialty was money laundering, a very useful and noble enterprise at that time.

We kept busy with a range of clients including the Reagan Administration and its involvement in the Contras' activities in Nicaragua.

An interesting career for a young prospector from Kalgoorlie!

However, all that came to an end when the managing director of the bank was murdered on Australia Day in 1980.

He had his head blown off.

Up until then I had often felt that the excitement of merchant banking beat the hell out of converting steam-driven mine winders to electric, back in Kalgoorlie. However, once again, like converting steam-driven mine winders, my position was declared redundant.

I eventually made peace with the ATO. They tore up my files and I once again returned to the Australian workforce.

That timing was fairly good as gold had suddenly stuck its head up, peaking at US$860 in January 1980, as a result of the oil crisis and the Middle East OPEC members who were desperate to pump their surplus funds into anything real, such as gold, as they were becoming increasingly alarmed at the future of the U.S. dollar.[16] (

So, let's take a break from the personal reminiscences for a moment and look back at this tumultuous period of Western Australia's exploration history. To illustrate the difference between what I call the two booms, let's look at the way both were celebrated in retrospect (always a more accurate view of history; looking back).

Kambalda Nickel 25th Anniversary Event

On 15 September 1992, Western Mining Corporation staged a 25th anniversary event for its senior staff to reflect on the significance and the very substantial outcome of the Kambalda discovery. That was the 25th anniversary of the opening of the Kambalda nickel mine by the then State Premier, the Hon. David Brand.

It was my good fortune that my wife Jenny and I were invited to these celebrations. As someone once mentioned, although I had never worked for Western Mining, I always 'seemed to be there'.

John Oliver, my earlier School of Mines colleague, and I continued

---

[16]   Prospecting a Way of Life ... see *Heroic Misadventures,* pp. 189 – 196.

a strong friendship and I clearly recall, one Saturday afternoon in 1966, he telephoned me and said: "Come on Manners, let's get out to Kambalda. We have to make some important decisions". The 'we' was Western Mining Corporation. As John was the Resident Manager, he drove out to a very unimpressive sheep watering trough at Kambalda and the two of us walked up the side of a hill overlooking the lake.

John said, "We have to start mining quickly here otherwise we will lose this nickel market opportunity. We haven't done enough drilling to properly decide where we should sink the shaft, so how about we simply sink it here because at least it has a good view of the lake?"

My vote didn't really count, but I agreed.

John then said, "What colour is the lake?"

I said, "It really wasn't a colour it was just simply silver."

"Okay, let's call it the Silver Lake Shaft," John said.

That's how decisions were made in those days before our industry became infested with committees. That's when managers were allowed to manage.

John was good at making decisions and I think he had named all the Kambalda streets before a Town Planning Committee was formed too. Again, that's the way decisions were made in those days before our industry surrendered to the various branches of bureaucracy.

Just look at this achievement timeline for the Kambalda operations:

- 28 January 1966 – Massive nickel sulphides were intersected at Kambalda.

- 21 February 1966 – The discovery was announced to the Australian Stock Exchange.

- 4 April 1966 – W.M. Morgan, Western Mining's M.D., announced a 'significant' discovery at Kambalda.

- 7 April 1966 – Announcement that 'under-ground development' would begin.

The key significant figure, of course, is that from the discovery hole until concentrates were produced was only 17 months. Just try achieving that today!

Yes, there was a lot to celebrate at the 25th anniversary. There, at that anniversary, they were celebrating a 100 per cent-owned productive asset, brought from discovery to production in record time by real leaders. Those were the days when miners were heroes, enthusiastically supported by communities, politicians and regulators, before the rot set in.

The Poseidon 20th Anniversary

I know about Poseidon's 20th reunion because I organised it.

Early in 1990 I had an idea. Should we allow the 20th anniversary of the Poseidon boom (the stock market peak of the Poseidon shares) to pass without a celebration of some kind? I felt we should gather together about 15 of the original Poseidon participants as it would guarantee a good night of reminiscences at Kalgoorlie's Hannan's Club. So, out went 15 faxed invitations. Instantaneously I received 52 acceptances, as the invitations had been passed on.

It resulted in a great night at the Hannan's Club, a star-studded evening of good humour, where, at last, many true stories of those times were revealed. In attendance were Norm Shierlaw, other Poseidon former directors and an incredible supporting cast. The event was ably reported by finance/resources writers Ross Louthean and Trevor Sykes whose stories were incorporated into a commemorative souvenir and it formed an airline magazine article, both of which were reproduced in my *Heroic Misadventures* book.[17]

The event was conducted in complete contrast to the Western Mining Kambalda event, but was nevertheless significant in its own way.

The nickel industry had quite a few tough years in the late 1970s and early 1980s, although our family company kept busy by introducing some productive tools, particularly in the form of the Wagner Scooptram diesel loaders from the U.S. and Kiruna low-profile underground trucks that we were importing from Sweden.

Then the region experienced a new dimension when gold started sending out a few green shoots following President Nixon's breaking of the link between gold and the U.S. dollar in 1971.

As I mentioned earlier, gold really got some wind into its sails with the massive investment from the Middle East oil producers turning their surplus cash into gold. This brought on a whole range of new listed companies.

There was a huge difference in this generation of gold companies compared with the earlier listed nickel companies, for two main reasons. One was to do with the tax implications. With nickel, there was a tax penalty for being on the Board of your company. However, with gold it was acceptable, so that's when we all became very 'hands on'. The second was that there was now an abundance of skilled technical and corporate people to go on to Boards (in complete contrast to the earlier nickel companies).

Among the companies that developed some substance were Metana Minerals, Sons of Gwalia, Delta Gold Ltd, Normandy, Great Central NL and Croesus Mining NL.

I'd like to mention a few highlights and lowlights of my time with Croesus Mining as I'm certainly more familiar with that company, although I was also a founding director of Great Central Mines NL.

---

17   *Heroic Misadventures – pages 290 and 294*

Croesus Mining NL – Highs and Lows

I do cover the 20-year lifespan of Croesus Mining in my *Heroic Misadventures* book, so I'll only touch on a few points here.

My other old School of Mines colleague, Keith Parry (Western Mining Corp.), mentioned to me, in the early 1980s, that I should stop joint venturing my mining properties out to other companies and actually form my own company to do all the exploration myself. Keith said, "If you get the right people together, you could be just like Western Mining Corporation before we lost our way."

Well, I did register Croesus Mining NL in 1982 and we were ready to move toward listing with some extensive properties at Polar Bear Peninsular near Norseman and just about all the Mount Monger Goldfield included. Well-known mining solicitor, Chris Lalor, agreed to be the Chairman.

We were moving toward listing when a former non-performing joint venture partner, White Industries Ltd, announced that it intended to exercise its 'pre-emptive right' over the property because once it knew what we were doing with it, it thought they should do the same and were moving toward putting the property into its own float, A.U.R. NL. I had several robust phone discussions with Travers Duncan. (White Industries) where I politely mentioned that White Industries had relinquished its interest and had never had a 'pre-emptive right'. However, they responded that they had their own in-house solicitors and they had more money than me and, from this, I knew that resolution would probably be some time away. So, I moved the Mount Monger areas into another public company called Mistral Mines NL and became its Exploration Director, until such time as I had gathered together the replacement land package for Croesus Mining. This took until late 1985 and by that time Chris Lalor was thoroughly occupied with his new role in Sons of Gwalia and I replaced him – albeit as Executive Chairman.

We listed in July 1986, the only company to float in that miserable year, with three well-known brokerage houses backing the float. Just for interest let me mention who they were:

- T.C. Coombs of London as underwriters.

- May & Mellor, leading Melbourne stockbrokers.

- Ray Porter & Partners from the Perth Stock Exchange (which was quite separate from the other Australian Stock Exchanges at that time).

It's interesting to reflect that none of these names exist today – an illustration of how volatile many aspects of our industry have been and will remain.

We raised $2m which we promptly spent on enthusiastic exploration which, to our disappointment, produced no mineable deposits anywhere.

At that point I realised the need for a cash flow (after many years in business I knew the importance of cash flow). I saw an opportunity when CRA (Rio Tinto) announced that their Goldfields Region prospects were all for sale as they had decided to move out of gold. I saw this as a great opportunity for Croesus and promptly headed to Perth for CRA's head office to ensure that we were on the tender list.

They and their advising bankers would not include us as we had no credit rating. So, I walked up St Georges Terrace to Westpac Bank and explained that we could not get on the tender list without a Letter of Credit from Westpac. Westpac, realistically, explained why they had some difficulty in issuing a company, who had just run out of money, with a Letter of Credit. With tongue in cheek I asked how much they would charge to actually *lend* me a Letter of Credit until noon. They hurriedly had a meeting in the next room where they probably had a good chuckle and returned to advise that they would charge me $25,000 to lend me a Letter of Credit until noon.

That's all I needed. I marched up to CRA's office proudly displaying this document and was consequently included on the tender list. They alarmed me by wishing to keep the Letter of Credit and I explained that it was mine and not theirs, but they should feel free to keep a photocopy of it. They found this to be satisfactory and I managed to get the Letter of Credit back to Westpac at one minute to noon!

Our tender offer for the CRA areas was $20.3m or 2.5 per cent higher than the highest bid, whichever was the higher figure. They pointed out that this was a non-conforming bid, but I managed to convince their banking advisors that their prime responsibility was to obtain the highest figure possible for their clients, and on that basis, they should honour that commitment. At the conclusion of the tendering process we walked away with that prize which, in my mind, mainly consisted of the Hannan South Mine.

There is quite a story on how we managed to pay the 10 per cent deposit by 10.00 a.m. the morning after winning the tender and finance the entire purchase within the 21-day prescribed period and managed to repay all the loans from gold production within nine months.

With much help from an enthusiastic board, staff and well-chosen consultants, with special mention of John B. Oliver and H.M. (Harry) Kitson, Croesus Mining NL embarked on a productive and profitable 20-year adventure where we mined 25 open pit and two underground gold mines and paid 11 dividends to our supportive shareholders.

The ill-advised so-called Native Title Act came into being as we were developing the Binduli mining area just south-west of Kalgoorlie and we ended up with eight competing claimants over that area at a time when we were all ready to build a new mill and feed it from several open pits. The eight claimants were basically all members of the same family and I could already see the nonsensical frustration that lay ahead of our industry.

Our mill never got built. The plans are still in plan cabinets. So, to stay in business we high-graded the mine and trucked the ore to our Hannan South Gold Operations. I estimate that that Native Title delays cost Croesus Mining $26m and that, to us, was a lot of money. The money didn't go to anyone, it was just that the gold was never produced.

I was not enjoying these non-mining frustrations that were encroaching on our industry and at that point decided to sell my interest in Croesus Mining. An interesting process emerged to find the most suitable incoming major controlling shareholder and our advisors had identified eight suitable companies.

This was a most interesting and complex procedure but one that I would recommend to any major shareholder planning to depart a public company. Somewhat similar to 'planning your own corporate funeral'.[18]

I brought senior staff into my confidence throughout this entire process and I was asked what I would do if I found that the best offer, to guarantee the future of the company and the fantastic team we had put together, was not found to be the same offer that gave me the highest dollar amount.

This, in fact, was the case and it actually cost me a quarter of a million dollars, but I felt the outcome was worthwhile. The incoming Canadian company, Eldorado Gold, was really acquiring a first-class technical team that they wished to deploy to Indonesia and participate in some new generation discoveries, i.e., Bre-X (that is a separate story for later).

Part of my contract with the incoming Eldorado Gold was that I would go on their Board and would cease being Croesus

---

[18]   Full story at www.mannwest.com

Executive Chairman within six months of the transaction and leave the company within two years. The events that unfolded made this extremely difficult. However, I had already moved out of Kalgoorlie to Perth and reactivated my family company which for several years been capably operated by my co-director, Harry M. Kitson. (See the following chapter 10 for more details of Croesus Mining NL and the part it played in Australia's Great Gold Renaissance.)

Following a series of tragedies, Norseman death, gold stealing charge and a decline in respect and mutual trust.

The eleventh and last dividend that Croesus paid was rammed through by me, much to the protests of senior staff. I felt, at that stage, that the cash would be more usefully utilised by the shareholders rather than left at the disposal of the new crop of executives.

Shortly after that I resigned as Chairman and quite a few months later a most efficient and effective period of administration was facilitated by Pitcher Partners who cleaned the company up, appointed a new Board with new properties and proceeded to turn it into something of an example of how intelligent exploration can create a new future for old companies and give former shareholders something to smile about.

A Quick View of a Patient Prospect

Polar Bear Prospect – Norseman

One of the properties maintained through this entire era; almost 40 years, was the original Polar Bear tenements and I'll just mention this because, again, it shows how patience can pay off.

In 1979 I was drawn to the area because of the early 1970s exploration by the U.S. company, Anaconda Inc, then managed by geologist Tony Hall, who later became a director of Croesus Mining NL.

An exploration program was commenced by Belgian company Union Minière.

Further exploration was later conducted by another U.S. company, Duval Mining.

My files show that following my pegging as Mannkal Mining Pty Ltd, the ground has passed through a series of companies' hands, including:

- Terrex Resources
- Croesus Mining
- White Industries
- A.U.R. NL
- Mareeba Mining
- Thames Mining
- Dominion Mining
- Sirius Resources

After some timid exploration, from so many companies, the ground came back to Croesus / Sirius / S2R

and now, 36 years after my original pegging, someone actually had the courage to drill a hole and bingo! Or, as geological writer, Keith Goode, said in his report – "Sirius hit the jackpot". Again, courage and imagination are reactivating our Australian exploration industry. Patience and perseverance have prevailed. Congratulations to Mark Bennett and his team at Sirius Resources.

How we lost our comparative advantage

A question: Where were our leaders and philosophers, when we had a chance to set some sensible rules on how the game should be played?

Nobel Laureate (1969) Paul Samuelson was once challenged by the mathematician Stanislaw Ulam to name one proposition in all of the social sciences which is both true and non-trivial. It was several years later that he thought of the correct response: comparative advantage. "That it is logically true need not be argued before a mathematician; that it is not trivial is attested by the thousands of important and intelligent men who have never been able to grasp the doctrine for themselves or to believe it after it was explained to them".[19]

I humbly suspect that our mining industry's bright and intelligent people were among those who didn't appreciate this concept of comparative advantage, otherwise they would have resisted repeated successful attempts by various enemies of our industry to de-knacker it. This has led to a decline from hero status to pathetic performance status, where investment returns are such that no-one in their right mind would rely on dividend flow from our resource companies to finance their future.

If what I'm saying is true, then this reflects badly on our industry leadership.

Talking about the word 'leadership', our Mannkal Foundation ran an essay contest (2010) at Curtin University's WA School of Mines. The contest challenged students to search for leadership in our industry and report on any examples they could find. There were 35 finalists in the contest. Overall it confirmed that, apart from the courageous few (three or four), there exists a leadership crisis in our industry – generally populated by puppets and caretakers.

The overuse of the word 'leadership' could lead to it becoming yet another weasel word where any real meaning is lost.

Here are some other words and descriptions that fall into the

---

[19] P.A. Samuelson, "The Way of an Economist." In *International Economic Relations: Proceedings of the Third Congress of the International Economic Association.* (London: MacMillan, 1969), pp.1-11.

category of weasel words or phrases:

- Sustainability
- Corporate Social Responsibility
- Social Justice
- Global Warming
- Political Correctness
- Native Title
- Stakeholders, where companies are ranking stakeholder interests before shareholder interest – it is no wonder that investors have gone elsewhere.

All of these words or phrases have been used by various rent-seekers as weapons in their campaigns. As Public Choice theory economists have shown, there is a lot of money to be made by the few who benefit, while the costs are designed to be spread across such a wide number of victims that, although debilitating, they are not life-threatening.

The result is that those who benefit greatly from these programs are incentivised to put in a huge effort to advance their interests, whereas the many who share the burden are not as well organised and continue to carry the financial burden.

The usefulness of Public Choice Theory in explaining bad government (or company) policies was explained to me by Professor James Buchanan in Moscow, September 1990 (as described in chapter 11). Robert Nozick described resulting government policy: "The illegitimate use of a State, by economic interests, for their own ends, is based upon pre-existing power of the State to enrich some at the expense of others."

Governments and corporate executives often give away money that is simply not theirs to give. If they had any understanding of Public Choice Theory, they would realise that the moral thing to do is simply say no to all these competing demands for corporate support.

Might I suggest that for every social style of seminar that mining executives attend, they should attend two seminars on becoming profitable.

Fortunately, the next generation is not hood-winked by this redistributionist nonsense. They are more anxious about the average debt of $85,000 that our generation has left for them to pay off (and that's not taking into account any student loans they might have).[20] It is 'their' money that the politicians are spending in order to be re-elected. This broken political system is seen clearly by the next generation and they are increasingly making their feelings obvious via their effective use of social media.

While on the subject of weasel words and nonsense served up to us in the media, do you remember the nonsense about the Rudd Government successfully steering Australia through the Global Financial Crisis (GFC) by giving Australians (alive or dead) a A$900 stimulus package with which to purchase an imported flat screen TV?

Yes, that explanation was served up with a straight face. The actual truth, far removed, was that the Campbell Committee forced our Australian banks to put their houses in order (around the time of the Asian Financial Crisis in 1996) and our banks hadn't had time to get themselves untidy again when the GFC hit. That's the real story of why we survived the GFC.

When did the tide start turning against our mining industry?

Perhaps some of you can remember the dark days of Gough Whitlam and Rex Connor in the mid-'70s, with the mass exodus of geologists overseas. Signs on office walls displayed the message: "Will the last businessman leaving Australia please turn off the lights."

We haven't ever recovered from that anti-development mindset. Sir Arvi Parbo concisely summed up Australia's business environment

---

[20] See 'Will they make the same mistake – Mannerism by Ron Mannes, 21-12-11 – www.mannwest.com

in an interview in the October 1995 edition of *Director* magazine:

> Today when you do something you know that from the first day probably half of the country is working against you in some way or another. Half of the government is working against you. You will have departments in favour of what you are doing, and probably an even number of departments very much against it. They will want to hem you in and stop you from what you are doing, or at the very least, make sure that you can only do it in a restricted manner.

That's when the leadership of our industry should have taken up Sir Arvi's challenge and instituted a set of rules for our industry that would have restored our comparative advantage. Now in 2015, it's late, but not too late, to do something about it. However, is anyone doing anything?

Well, last week I saw one example and that is Gina Rinehart. She meticulously documented the 4,940 approvals, permits and licences that they have so far been forced to comply with to bring in their single Roy Hill mine. This list was submitted to Premier Colin Barnett for his consideration and comment. Let's wait anxiously for his response. (The letter is included at the end of this chapter.)

Why can't we see more leadership like this in our industry? Where are our mining leaders when we need them?

I always enjoy going through my old files and over the weekend I found some notes from a speech I gave in Toronto at the Prospectors and Developers Association of Canada (PDAC) in March 1977.

Amongst the things I mentioned in Canada was the reason why Australians and their money were leaving Australia. It had simply become too difficult for us to get on with the job. I also described some differences between Australians and Canadians and generalised by saying that Australian's take business very seriously but at the same time don't take themselves very seriously.

Once Canadians realised that comfortable contradiction, I continued, they will understand the Australians and their peculiar sense of humour. For example, it probably explains why male Australian geologists always give their penis a name. They can't stand the thought of their major decisions in life being made by a stranger!

Back to reality. While we are on the subject of regulations, rules and general impediments to industry, I should caution about adopting a victimhood status for our industry. Why? Well, victims cannot be leaders.

If you view yourselves as a victim in any corporate sense, or in any aspect of your own life, you are giving your power away. If you are a victim, someone else has to change to make you happy but you cannot change anyone else.

For leadership to emerge from our industry, we must get up off our knees and change ourselves. It's part of the individual responsibility package that comes with true leadership.

This victimhood angle is one of the reasons why the Native Title Act fiasco has failed the aboriginal people. From their perspective it has been based entirely on the cult of victimhood.

My extensive files go right back to the beginning of the Native Title fiasco. I recommend two books: *Red over Black* by Geoff McDonald (1982) and *The Fabrication of Aboriginal History* by Keith Windschuttle, Volume One, Van Diemen's Land 1803–1847 (published in 2002). I have written much on the absurdities of the Native Title Act and in 2004, the last time I calculated native title cost of lost production to Australia, it was $90 billion.[21] All this and it gives no title to our indigenous people.

Just before leaving the topic of leadership, let me mention several

---

[21]   See *Heroic Misadventures* – Dubious Land Title (Land access & Property Rights) pp. 104-107. *Never A Dull Moment* – p. 424.

specific examples of how we are failing:

1. In Hong Kong, at the big Rugby 7s 40th Commemorative Dinner – 2016, I heard two rugby legends each speak for 40 minutes. They both spoke without notes and with great passion. I sat there wondering why our industry, the most imaginative and creative industry in the world, has failed to create a significant collection of people who can speak with similar passion about their industry?

2. In 2015, we witnessed two political funerals (Malcolm Fraser and Lee Kuan Yew) but when our own industry legends pass away there is hardly a murmur in the media.

I mentioned Norm Shierlaw as the father of the Poseidon boom. When Norm died in September 2013, I didn't hear about it until April 2014 (six months later when I read his obituary in the AusIMM *Bulletin*).

When Kingsgate's Gavin Thomas died in June 2014, apart from his company's notice of his memorial service, there appeared to be no mention in the media.

When Geoff Donaldson died in July 2013 – again, hardly a media mention. Without Geoff Donaldson there would be no Woodside Petroleum.

We have lost our ability to project our industry as an exciting, creative and profitable endeavour to the point of encouraging investors to join us in our risk-taking. If our industry is not prepared to take the risks themselves, then they must accept the fact that the acknowledgement prizes will not go to them either.

Wisdom and Data

So, we come to the final question raised earlier: What are we going to do with all the wisdom and all that data that we have accumulated over all these wonderful 50 years?

I'll bet everyone in this room constantly receives requests for information on the very substantial areas with which you have been involved.

Recently, I received two such requests. One from Rohan Williams (now at Dacian Gold Ltd), the new owner of one of Croesus' 1986 mines, Jupiter, at Laverton and the second from Colin McIntyre, now responsible for breathing some life into Norseman.

Colin was looking for a 2003 study we completed on several open pit optimisations, including HV5B pit (on the edge of Lake Cowan), when the gold price was A$540. Now (2015), at a current gold price of A$1,500, those pits start to look robust.

Never underestimate the value of your old data, as it will be treasured by those who follow. Equally important is the need to make yourselves available to the new and important generation of explorers.

If you haven't found a home for your own data, let me encourage you to do so and do it quickly, before your descendants mistake it for garbage and send it off accordingly. I had about 2.5 tonnes of such material and it wasn't easy giving it away.

Nobody wanted 'the lot', so, in 2006, after much discussion, I split it into three specific packages (they're all about the same weight) and I'll mention this in case it is useful guidance for you.

One third – advanced prospects, that could emerge when circumstances were right – went to a young entrepreneurial mining engineer and I have the opportunity to contribute to any commercial development that may eventuate.

One-third went to Curtin University's School of Mines – an extensive collection of maps, aerial photos, sundry papers and reports. This was all co-ordinated by Professor Philip Maxwell, Bob Fagan, Kerry Bradford and Libero Parisotto.

The remaining third then went to the Geological Survey of W.A., an extensive collection of company prospectuses and annual reports covering the 50-year period, along with a laptop and database. You might remember that a company prospectus of that era contained considerable useful information. They contained a bibliography which listed all earlier reports and references on the ground covered by that company's tenements.

I had databased all this company information, cross-referencing to the subject mineral, authors, companies and other details. This information was well received, because some time before that, geologist Bob Pickering had left his collection to the Geological Survey, but they had misplaced his database and indicated that Bob's information could easily be included on my database.

This information hand-over was co-ordinated by Dr Ian Ruddock and Tim Griffin and in their acknowledging letter on 4 April 2006 they stated:

> The collection will mainly be used by our Mineral Resources Group as a useful source of data for the WAMIN and MINEEX data bases on mines, deposits, prospects and occurrences. It would also be made available to the public and the Library at Mineral House.

So, I was then 2.5 tonnes lighter and could once again fit my car into our garage. I hope my old data, along with your own, will guide future generations towards some exciting discoveries.

Let's finish with some good news (particularly for gold).

Recently, in Hong Kong, I had lunch with Doug Casey, the noted investment commentator.

He showed me his useful gold graph. It indicated that the current downturn in the U.S. dollars gold price is not much different to previous percentage downturns and as we both have tremendous confidence in governments' ability to destroy currencies there is certainly a great future for gold.

Conclusion

Friends, it has been my pleasure to join this distinguished group of active participants in this fast and furious 50 years' retrospective. They say that all progress takes place outside the comfort zone and this is where your achievements were made. The words of the old American Indian proverb could have been written with you all in mind:

"You will be known forever, by the tracks you leave."

Email sent to: Hon. Colin Barnett, Premier of Western Australia

From: Mrs Gina Rinehart

Dear Colin,

Thank you for your letter of 6 February. Good to see you tonight.

My reply has been delayed as my team dealing with approvals, licences and permits, tried to include a register of same that the contractors and subcontractors had to achieve to be able to construct Roy Hill. I understand they've tallied around an additional 800 so far, making a total of 4,940, but many more need to be included. I've asked Bill Hart to attach these lists. However I didn't want to delay further given Barry Fitzgerald, of Roy Hill, will be referring to this also in his presentation at a conference this Wednesday in Perth.

I asked various political people present tonight how many government approvals, permits licences they believed were required for Roy Hill, and received the usual responses, "several hundred". And the usual responses when I tell them no, it is more than 4000 and that doesn't include the many hundreds or thousands more for actual construction. To a one, they were shocked. Colin it's not the individual ones that cause the most problems, it's that there are thousands too many.

One can and should ask, where the next definite projects in WA's pipeline will come from after Roy Hill, to maintain our pride in W.A's impressive record of achieving 50 percent of Australia's export income and around 16 percent of our country's GDP. Even Woodside has baulked at the thousands of approvals needed for on-land W.A., some further requiring thousands of responses from any member of the public, when only about 5 are required if they conduct those operations offshore. We are simply not going to maintain W.A's export and GDP position via nice beaches and wine. Dealing with such government burdens puts a real extra pressure also on Australia's declining productivity and of course Australia's very high costs.

In my view, the cost of government must be significantly and urgently reduced, as we face record debt in West Australia and across the country, and importantly the burdens on business via government, and its approvals, licences, permits and compliance costs must also be urgently and significantly reduced.

I hope we can find common ground in encouraging significant and urgent change in these areas.

All best,

Gina Rinehart

SHORT-TERM LETTER OF CREDIT

John Dow, President Australian Gold Council
Presenting the award to Croesus Mining, Australian Gold Company of the Year - 2003

9

CROESUS MINING

This is a story of gold mining. My family had been in Kalgoorlie for three generations, designing and building mines, converting steam-driven mine winders to electric and providing a wide range of mining equipment, so mining was in my blood. My great-grandfather, in Ballarat, Victoria, had taken this process right through from prospecting to production and I had always wanted to do the same in Kalgoorlie.

Despite today's misguided perception of 'mining', this sequence of prospecting to production is a highly creative process. I recall seeing a sign from the wall of a Peruvian mine site that captures this process perfectly.

> It is important to eradicate the widely spread concept that finding a mine is like discovering a treasure. A mine is not a hill where there are minerals. A mine is the effort of individuals who have converted the hill into a fountain of labour and of wealth. It is a product of sacrifice, of intelligence and technology, of management ability, and of extraordinary tenacity.

I will confine this story to my 20 years with Croesus Mining NL (1985–2005), during which time I often reminded our great team that "we are not in the exploration and mining business, which simply describes what we do. We are in the business of turning ideas into gold bars, profitably".

The story really started several years earlier. During the early 1970s, the Australian Taxation Office (ATO) claimed that I owed tax on some mining vendor shares in Westralian Nickel NL on the basis of their value at a time when their market price was $8 each. They were subject to a 12-month escrow, and so could not be sold until the period of "escrow" expired. By that time the nickel boom was over, and I sold them for 15¢ each! However, the ATO felt that this drop in value was irrelevant (they actually said it was 'my problem') and continued its efforts to extract from me this fictitious sum of money, on which they were charging 10 per cent interest which, annually, was more than my salary!

Despite my many requests that they 'take me off their mailing list', they did not oblige. About the only polite letter I got from them during the next seven years was one announcing that the interest rate had been raised from 10 per cent to 20 per cent! It was during that time that I adopted the motto: "If socialism works, I won't", I dropped out of the productive workforce and enjoyed what I will probably come to regard as the most interesting period in my business life.

My time was busily spent in writing letters to various editors, forming a new political party and travelling overseas, where I ran hotels, explored for nickel and wrote a story entitled *The Alienated Australians* (see Chapter 6). For this story, I interviewed 27 other Australians who had been blown out of the Australian workforce by the heavy hand of the bureaucracy. They were all enjoying successful careers in other countries where they were able to move so much faster, without the bureaucratic burdens of Australia. (Now, in 2019, I see this situation returning.)

However, all good things come to an end and I started to feel that I really had to have a proper job, particularly as it was embarrassing for my children when they were asked: "But what does your father really do?" I also realised that my motto: "If socialism works, I won't", just didn't make sense as socialism has never ever worked.

In the end, socialism just runs out of victims.

After about seven years of 'working offshore', I wrote politely to the ATO (in the early 1980s) and suggested that if they tore up all their previous files[22], I would be prepared to re-enter the productive workforce and allow them to pursue yet another pay packet. They found this offer too good to refuse and this gave me the opportunity to re-activate some earlier skills in floating a listed mining company which I christened "Croesus Mining".

Why pick a name like "Croesus", a word that very few could pronounce let alone spell? Croesus was the King of Lydia, now Western Turkey, between 560–546 BC. He was noted for his great wealth, hence the term "rich as Croesus". Anywhere in the world where there is mining, you are likely to find a Croesus Hill or a Croesus Nob or a Croesus Shaft, and in Kalgoorlie there is a Croesus Street where I lived and provided the company with a registered office. (The operating staff were based in our home for the first couple of years of the company's life.)

King Croesus, as a means of survival through good and bad times, formed many joint-ventures with neighbouring countries in much the same way as I had been joint-venturing my mining properties. This gives tremendous leverage. By doing deals with many of these partners, he raised Lydia to the peak of its power, conquering Greek coastal cities and extending his empire to the Halys River in Central Asia Minor.

As King Croesus is also famous for having been the first ruler in the world ever to mint gold coins, it seemed a good name to use!

Although I registered the company in 1982, it was four years later that the company listed. My many exploration projects kept producing results requiring even more exploration, finding their way into already listed companies where I became Exploration Director/

---

[22] It was not quite that simple. I owe considerable thanks to U.S. Attorney, Clyde R. Maxwell, who convinced the ATO that "In any civilised country if the taxation legislation is ambiguous it must always be interpreted on the side of the taxpayer.

Manager or a similar role.

By 1986 I had a group of prospects that I managed to 'keep still' long enough to assemble a prospectus for a 'public float' (now termed an IPO).[23] 1986 was a low point for the industry and Croesus was the solitary gold company that listed that year.

Croesus Mining began trading on the Australian Stock Exchange on 24 July 1986 having raised $2.5 million. After nine months of unsuccessful exploring, we were down to $0.5 million with no sign of any cash-flow. We needed to be bold and act quickly. The result was that, a little surprisingly to us and others, we acquired Rio Tinto's[24] gold interests in the Kalgoorlie region for $20.3 million. We were attracted to Rio's recent gold discovery of easily mineable gold right alongside our own exploration tenements.

Rio Tinto would not initially talk to us in the tender process because we had no 'credit rating' and they insisted that we produce a Letter of Credit from our bankers. Our bankers, Westpac, were understandably unwilling to give us a Letter of Credit for $20 million! I asked how much Westpac would charge if we only wanted to 'borrow' a Letter of Credit for the morning. Their answer was $25,000 and I had to promise to have it back to them by noon.

This enabled Croesus to be included on Rio Tinto's tender list, but it was a bit embarrassing when Rio Tinto asked to keep the Letter of Credit. They eventually settled for keeping a photocopy and I managed to get the original back to Westpac at one minute to noon.

Looking back now, at this highly unusual provision of such a 'facility', Westpac must have felt this was an easy $25,000 for providing me with this document with which I could not have done much 'damage'!

John Oliver, a friend since our time together at the Kalgoorlie School of Mines in the early 1960s, and who had co-ordinated

---

23 Initial Public Offering.
24 Then called Conzinc Riotinto of Australia Ltd – CRA.

Western Mining Corporation's Kambalda nickel operations right through to production, was my mining adviser and he thought that a bid of $20.3 million was justified.

It happened that John had good connections with Bank National de Paris (BNP), and we brought this bank's senior executive from Sydney to Kalgoorlie to look over our proposal. They agreed to finance us. BNP took a copy of all the data and dispatched it to their head office in Paris where they confirmed that, technically, our courageous bid was sound.

Subsequently, our bid won the tender and later that day I phoned BNP in Sydney (just before closing time). Their Australian manager said he would immediately telephone their Paris head office which would just be starting their day. He phoned me back saying that they were delighted and that their mining engineer would be visiting Australia in about nine months, at which time they would then be in a position to discharge funds. As we had to have the 10 per cent deposit ($2.03 million) into Rio Tinto by noon the next day (remembering we only had $0.5 million left ourselves), with 21 days to pay the balance, I could see that BNP would not work for us.

My main concern, at that moment, was not only about missing out on the deal but also how I was to explain to the auditors the $25,000 payment to 'borrow' a Letter of Credit.

A few hours later, I received an abusive telephone call from London. Rod Whyte, a stockbroker friend of mine had read on the Reuters' wire the details of the transaction and was upset at not being given the opportunity to participate. I said, "Don't worry, there is still time. If you place some Croesus shares right now with your UK clients, and if you promise to have $2.5 million deposited in our bank account by 10.00 a.m. tomorrow morning, you can be part of this exciting story."

Thankfully this worked and getting the other $18 million together over the next 21 days, in a mixture of share placements and loans, was equally exciting. Only with a lot of help from some very supportive friends and colleagues were we able to get the transaction over the line.

My father always told me, "Never borrow money to bet on horses or look for gold", but I felt comfortable about borrowing money against the proven gold reserves actually in the ground. Those reserves were well in excess of what we needed to pay off all the debt within nine months through a combination of mining that gold and embarking upon various exploration programs. In total, we established 25 surface and two underground mines over the next 14 years.

Then, in 2002, Croesus jumped a few similar hurdles when we took over Central Norseman Gold Corporation (from WMC Resources – Western Mining Corporation). Our bankers asked if we were really serious when we set out to borrow $60 million which, they reminded us, was "twice your current market capitalization". We raised the funds, repaid all debt and emerged with a market capitalization of $150 million.

By 2005, Croesus employed about 420 people and had paid 11 dividends, totalling $28 million, as a result of mining and pouring 1.275 million ounces of gold which, at recent values, would have a total value of over $2 billion.

Mission Drift

Now, all these 20 years with Croesus Mining look fairly well organised, but the reality was somewhat different. Many gut-wrenching moments occurred during the later years at Croesus and now, upon reflection, these stories can be told.

By 1995, some of the difficulties for the mining industry caused by the Native Title Act were becoming apparent. It certainly gave us some headaches. Even after obtaining all the necessary approvals, agreements and licences to build a new processing plant on site at Binduli,[25] we were still unable to proceed due to the multiple

---

[25] 8 kilometres south of Kalgoorlie.

(eight) Native Title claimants. We were forced to 'high grade'[26] these deposits and truck the ore to our existing treatment plant, some 25 kilometres away. The unfortunate aspect of this was that we had to leave many thousands of tonnes of lower grade material behind, perhaps never to be treated.

After several years of intense frustration, I avoided attendance at any so-called 'negotiations', for fear of 'blowing up the ship'. (About that time, I was quoted in the media as saying: "Political correctness, this arrogant assault on plain forthright communication, is seen by them (leaders of big companies) to be an insurmountable barrier that prevents them from speaking their mind.") I realised this would be an appropriate time for me to hand over our successful and profitable company to 'others'.

Over a period of some six months during 1996, we managed the 'hand-over project' as one would conduct a job interview for an incoming Chief Executive. There were nine expressions of interest from other mining companies for us to evaluate. We insisted that directors and senior management of the interested parties visit us and that they be interviewed while they were evaluating our assets. We needed to know who we would be dealing with in the future.

The ultimate winner of this process was a successful Canadian-listed gold mining company, Eldorado Gold Ltd. It bought my 30 per cent shareholding in Croesus Mining NL (by acquiring my private company Mannkal Mining Pty Ltd). Eldorado was not deterred by the emerging Native Title problems in Australia as they, themselves, had extensive experience with similar legislation in Canada. This resulted in Eldorado not spending any exploration dollars in Canada, a very similar situation to that which now exists in Australia. The last time I investigated (2005) I found that 78 per cent of Australia's exploration budget was being spent outside of Australia, in Africa, South America, Eastern Europe and Indonesia.

---

[26] This meant that only the highest grade was treated, which reduced the mine life.

The hand-over appeared to have gone smoothly and two Eldorado Directors were appointed to the Croesus Mining Board and I was appointed to the Board of Eldorado. However, it wasn't long after the successful handover, that Eldorado was confronted with a savage sequence of external events.

At a recent long and pleasurable reunion lunch (March 2018) of the fantastic old 'Croesus team'[1], many of the events of those times were recounted. I realised then, for the first time, that it was not generally known to them that part of the hand-over documentation stated that my continued time as Chairman of Croesus was to be limited to six months. The assumption was that, by that time, Eldorado would be fully in charge and pursuing their stated intention, well known to our team, to embark on an aggressive exploration program in Indonesia. Eldorado's own shareholders were enthusiastically pushing them to replicate what appeared to be an amazing exploration success of the Canadian company Bre-X Minerals Ltd. In a blaze of publicity, Bre-X claimed to have discovered the world's largest gold deposit, in Indonesia (Busang in East Kalimantan).

This smooth transition was violently interrupted by two unexpected events in 1997, almost happening simultaneously. The first was that Eldorado had an opportunity to acquire major gold assets in Brazil, Turkey, and Chile from Gencor/Billiton of South Africa. This was good news for Croesus/Eldorado, with the acquisition being consummated at a very high-profile London event, presided over by Brian Gilbertson, then Gencor CEO. As a result of this acquisition, Gencor became a 40 per cent major shareholder of Eldorado and not long after it issued an edict that, as 'Gencor/Billiton' was already active in Indonesia, they didn't want Eldorado duplicating Gencor's activity by having any presence there.

This negated the prime purpose for Eldorado's entry into Croesus Mining which was that Eldorado was planning to have our exploration team lead their entry into Indonesia. Meanwhile, my monthly flights from Kalgoorlie to Vancouver continued, attending the Eldorado

Board meetings and international operations. However, it was very noticeable that the Eldorado directors were not intending to attend our Kalgoorlie-based Croesus Mining's Board meetings.

Then, at about that time, the 'fabulous Bre-X discovery in Indonesia', the "world-class discovery" was revealed to be, in fact, the mining industry's greatest fraud ever recorded. There was huge international interest in the events of those times, particularly when a Bre-X geologist was hurled from a helicopter by persons unknown. His body was ultimately recovered from the jungle but by that time had been eaten by wild pigs to the extent that identification was inconclusive. These events were incredibly damaging to the entire mining industry's reputation.

I happened to be very close to those events as I was at the presentation dinner where the President of Bre-X received the Prospectors and Developers Association of Canada's (PDAC) Award for 'Company of the Year'. As a guest speaker it was my 'honour' to be seated at the top table between the Bre-X President and the Exploration Director/Chief Geologist. Both these gentlemen were suddenly called to their hotel room. Their absence from the dinner table, mid-meal, was noted. We had no idea that they had been called upstairs to receive the news of the fraudulent 'discovery of the century'. This was revealed later.

However, they coolly returned to the dinner table and basked in the glory of their 'trophy award'. Their ultimate fate was subsequently detailed by the media.[27]

Eldorado's subsequent loss of interest in its Croesus investment resulted in it disposing of its 30 per cent shareholding to a collection of Australian-based shareholders with two of them holding major blocks of shares. I also repurchased a minor parcel of shares, as I

---

[27] For details of Bre-X, refer to the book *The Bre-X Fraud* by Douglas Goold and Andrew Wills. The spectacular events of that collapse were also covered in a film named "Gold", released in 2007.

was still impressed by the company's assets.

The agreement that I leave the Croesus Board within six months of 'hand-over' somehow became irrelevant. However, I had already commenced my move from Kalgoorlie to Perth. My family company, now based in Perth, needed my attention which was now possible as I no longer held any executive role at Croesus.

From this point on, the company became somewhat directionless as we were all waiting for one of the new major shareholders to step forward and 'put their mark' on the company and assume a managerial role. But this did not happen. Little did I realise that the 'nobody home' syndrome would continue for another five years, with the company being carried forward by its past momentum. During this time, executives came and went, none having any financial investment in the company or, as they call it, 'skin in the game'. Croesus was drifting. However, exploration and production continued and during the lead-up to this 'nobody home' status, Croesus became Australia's third-largest Australian-listed gold company (and for a short while the second largest when one of the other producers collapsed).

At its peak, in 2003, Croesus Mining received the award of Australian Gold Company of the Year.

At this time our staff and contractor numbers had increased to 420, from the original team of six that had effectively brought Croesus into production. There was a general sense of dysfunction appearing as a few personnel had been brought in from larger companies and they appeared to have no sense of 'keeping costs under control', as was very necessary in smaller companies. It wasn't long before a few things started going wrong, such as our Underground Superintendent committing suicide underground at Norseman, immediately followed by nine of our main underground miners being arrested for suspected gold stealing. That was tragic for a company that, for so long, had operated as an 'extended family'.

We knew that something was wrong at Norseman and it was more than obvious – such as feeding the township from the company dining room or providing so many children with free pens and stationery. The problems were so serious that we put one of Australia's best mine operators on site and he reported back to the Board. We concluded that the challenge of 'culture change' was significant and should be borne in mind when selecting Croesus' next Executive Chairman.

It was with some relief that I left the Board on 11 November 2005, though still as a shareholder, continuing my faith in the company's underlying assets.

Eight months later, the continuing Board enlisted the aid of an external Administrator with the announcement "Croesus Mining went into administration in June 2006 following operational difficulties at the company's prized Norseman mine".

It was agonising to watch the process from a distance. Fortunately, the administrator executed his role effectively, giving the continuing shareholders, including myself, a sporting chance as we emerged with shares firstly in Sirius[28] and then through subsequent corporate machinations, the original shareholders became shareholders in both Independence Group and S2 Resources.

And so, Croesus, the mining company, retired from the field, having led a very colourful life. After 20 years of dividends and gold production, the original Croesus shareholders are still hanging in there with a chance.

> "Discovery consists of seeing what everybody else has seen and thinking what nobody else has thought."
> – Albert von Szent-Gyorgyi

> "The cave you fear to enter holds the treasure that you seek."
> – Joseph Campbell

---

[28] Taken over by Independence Group NL 2015

TO *FREEDOM* СВОБОДЕ

TRANSITION ПУТЬ К

THE NEW SOVIET CHALLENGE
A Cato Institute Conference on the Changing Soviet System
Cosponsored with the Academy of People's Economy, the Academy
of Sciences of the USSR, the Central Economic-Mathematical
Institute, and Moscow State University

September 10–14, 1990   Moscow

Cato Institute
224 Second Street S.E.
Washington D.C. 20003
(202) 546-0200
Fax (202) 546-0728

Quantum Bureau
Academy of Sciences
of the USSR
Moscow V-71
Leninski Prospect, 14
231-8242
Fax 251-5557

Organizing Committee

Dr. Abel Aganbegyan
President
Academy of People's
Economy

Fyodor M. Burlatsky
Member of the
Supreme Soviet

Dr. Ted Galen Carpenter
Director of Foreign
Policy Studies
Cato Institute

Edward H. Crane
President
Cato Institute

Dr. Vladimir Dobronikov
Vice President
Moscow State University

Dr. James A. Dorn
Vice President for
Academic Affairs
Cato Institute

Dr. Sergei Kapitza
Chairman
Quantum Bureau
Academy of Sciences
of the USSR

Dr. Edward D. Lozansky
President
Independent University

Dr. Valery Makarov
Director
Central Economic-
Mathematical Institute
Academy of Sciences
of the USSR

Dr. Yuri Osipyan
Vice President
Academy of Sciences
of the USSR

CATO Institute's 'Transition to Freedom' team in Russia —
September 1990

An appreciative Moscow audience tunes in to the free-market —
September 1990

10

RUSSIA

'SEVEN DAYS THAT SHOOK THE WORLD'

(September 1990)

Life is full of invitations and one of the most interesting I have ever received was to visit Russia in September 1990. The event was a high-level strategic conference, jointly sponsored by Washington's CATO Institute and the Soviet Academy of Sciences.

Forty international academics, economists and businesspeople were to attend sessions in Moscow and Saint Petersburg to explain 'free-enterprise', a system that 'may be available to them soon'. It arrived that week.

Our timing could not have been better, although I'm sure that when putting the event together the organisers may not have realised that our visit would span the 'seven days that shook the world'.

We had ring-side seats to rioting in the streets, the dragging down of statues and the renaming of St. Petersburg (from Leningrad).

The proceedings were reported, in detail (see Appendix article by Paul Craig Roberts, *National Review – October 15 1990 –* 'Seven Days

That Shook the World').

Early in the conference, Ed Crane – President of CATO Institute – announced casually from the podium that, "The government that governs least, governs best", a line that all of us visitors had heard before. However, when the Russian interpreter finished translating it into Russian, the Russians leapt to their feet, erupting in wild applause. These words were encouraging to them, although there were quite a few expressionless Russian Commissars who showed little enthusiasm for this sentiment.

Our audiences consisted of high-ranking government officials and Communist Party operatives, intellectuals and sundry heroic dissidents. Some of the meetings numbered more than a thousand people, all leaning forward eagerly with their translating devices to their ears. Professor James Buchanan (Nobel Prize winner in Economic Sciences, 1986) was telling any Russian who cared to listen that one of the greatest dangers they faced was to have their economy captured by special interest groups. As a result of this they could end up with a catastrophe that would bear no resemblance to a free-market economy, indeed one that would deliver them absolutely none of the anticipated benefits.

He explained that a study of Public Choice theory would explain how political decision-making quite often results in outcomes that conflict with the preferences of the general public. Advocacy groups, with their pork-barrel projects, could be the result of the painful revolution they were currently experiencing. He painstakingly explained to them the extreme lengths to which those receiving the 'concentrated benefits' would go to bring about their objectives with the diffused costs spread over many millions of Russians who would be unaware of what was happening. (Buchanan would have used the term 'business oligarchs' if it existed at that time.)

Apart from the large audiences, there were many smaller seminars. At one of these seminars I was asked to explain how public companies were formed and capital was raised (a concept quite novel to the audience). To me this was an easy task, so I went through the steps of creating a business vision, then a business plan which could be presented to 500 potential investors (that being the required number for a public company). If you attracted 500 shareholders in this way, it was because they believed in your vision and felt that they had a chance to see their share value increase and dividends paid in due course.

Most of my presentation was received in stony-faced silence and at the end of the evening one large Russian gentleman came up to me, gripped me by the shoulders, and said "I don't know anyone in this world that I would trust with my money in this way!"

It was at about that point in the conference that I realised how difficult these concepts were for the Russians: they had no property rights, no law courts, no way of resolving disputes, in fact none of the 'building blocks' of a civilised society into which most of us, from the West, were fortunate enough to have been born.

As for day-to-day life in Russia at the time, we were well briefed before our visit and had some idea of what lay before us, although there were still some surprises.

The Russians had become so focused on just a few national priorities that they couldn't even make a ballpoint pen that worked, or a tube of toothpaste that didn't squirt out the sides.

We were even asked to bring our own toilet paper, as Russia couldn't even produce that in an acceptable quality. I even brought back a roll of Russian toilet paper and passed it around my office saying: "Russian toilet paper was like Communism itself. Full of gaping holes."

The 1990 health statistics were also alarming (for example, male deaths per 100,000 of population, caused by heart disease, comparing three countries: Japan 55, Australia 280 and Russia 480.) Despite the detailed briefing, nothing had prepared us for the experience of seeing our conference food being delivered to our hotel from trucks under armed guard. When I asked, "What's that all about?", I was told that the people in the streets were starving and that without the guards the food would be looted.

Looking back, I can say that the food wasn't worth guarding anyway and that led to three of us visiting downtown Moscow one night to see how the locals ate. Our experience was grim. Limited menus, mostly bowls of animal fat, although one of us did find a small potato lying at the very bottom. Russian food in 1990 probably explained the disturbing health statistics.

However, the dire nature of the food was more than compensated for by the calibre of the group of intellectual dissidents with whom we mixed. They included some of the bravest and most courageous people I have ever met. This was their opportunity to speak their minds and ask questions without being arrested. Many of these brave souls were mentioned in Professor Peter J. Boetke's 1993 book, *Why Perestroika Failed*.

One observation was that every Russian we met, appeared to have written one or more books, either literature or poetry. My colleague, Ron Kitching, often jokingly advised the Russians that I was the Australian poet of the group. I did not dare to inflict any of my Aussie doggerel on them.) I was inundated with so many poetry books that I asked one of the Russians why it was that they all appeared so much better educated than us visitors from 'the West'. To which he replied sadly, "When we complete our university studies, there are no jobs, so we repeatedly return to the university!"

Two other members of our delegation were greatly admired by these 'Russian heroes' because, for several years, the Russians had been receiving boxes of smuggled books from Tom Palmer who had arranged Russian translations. Similarly, boxes of the original English publications had been sent to them from Australia by Ron

Kitching. I could write a whole chapter on the various members of that 40-person team and how they have continued to add value to the fabric of civilisation in the ensuing 30 years.

George Gilder is one and I was reminded of this when I recently watched his YouTube – George Gilder, life after Google. Australia's Professor Ross Parish (1929–2001) was a great contributor due to his economic skills and knowledge of agricultural economics. His full report on the Russian visit (covering all aspects of the conference) will be available in the web-based appendix for this chapter. In addition, the paper by Professor Ralph Raico (1936–2016), "Liberation from the Parasite State", along with CATO President Ed Crane's much earlier 1982 paper, "Fear and Loathing in the Soviet Union". Ed's December 1981 visit sparked his interest in the Soviet Union and started him thinking of who he would include in his investigative team of 40. Ed's comments from his first trip were:

> I went to the Soviet Union in 1981 and was dumbfounded to see that, while the CIA was claiming the Soviets had 63% of our GDP, they, in fact, had no GDP. Little food, no consumer goods, lousy tanks and nothing else. (Okay some nukes.) But whatever GDP they had amounted to perhaps less than 5% of what we had in the US. I wrote an article in December of 1981 making that very point, predicting the imminent demise of the USSR.

I've told you about the background to the conference and given you a brief outline of our time in Moscow and St. Petersburg, but what are the long-term results of our efforts to help bring economic reforms to Russia?

At a 2005 economic conference in Iceland, I developed a friendship with Andrei Illarionov (Vladimir Putin's Chief Economic Advisor at that time) and he gave me some encouraging examples of how Russia had followed Estonia's example in adopting a flat rate tax (13 per cent in Russia). Andrei described this as "the best thing that Russia has ever done as Russians are actually paying their tax and on time too".

Andrei and I have met several times, since then, most recently in Texas in May 2019. He confirmed that the data he sent me in February 2019 was still current and proved, without doubt, that Russia exhibited a growing economy during the brief periods when the size of government shrunk but retreated each time government expanded. They had only managed a 2.3 per cent growth in GDP over the whole ten years from 2008–2018. "As government expands; economies collapse."

Andrei is no longer President's Putin's Economic Advisor and he is now a Senior Fellow at the Centre for Global Liberty and Prosperity at the CATO Institute in Washington DC.

The chart below, based on 2017 figures from *The Economist*, shows how Russia's economy is not dramatically larger than Spain's or Australia's. One might say that Russia is again creating its own myth. Russia is a regional power, pretending to be a world power, and getting away with it!

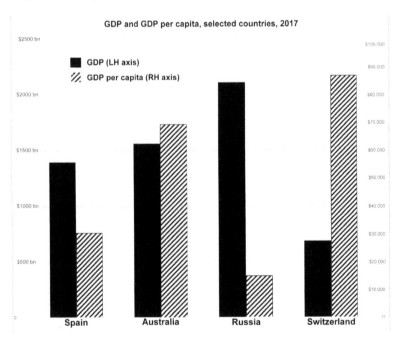

GDP and GDP per capita, selected countries, 2017

Andrei Illarionov and Ron Manners, Texas – May 2019

Memories of Moscow – September 1990

What an opportunity came my way in 1990. To be part of a bright bunch of 40 individuals with the challenge of preparing the citizens of Moscow and St. Petersburg for 'free-enterprise', which was arriving 'right now'.

I've already detailed this almost insurmountable challenge earlier in this chapter, but one less serious episode occurred when four delegates escaped for one afternoon to experience Moscow's street culture.

Joining me were Dr Hannes H Gissurarson (Iceland), Dr Roberto Salinas Leon (Mexico) and Ronald Kitching (Australia).

We had befriended a young Moscow university student, Alex Deinega, and he acted as our tour guide (also vigorously questioning us as he was intent on taking over the yet-to-be-privatised university coffee shop).

Part of our tour was the obligatory event of being sketched by one of the many sidewalk artists.

This was a very speedy process for my three colleagues who were handed a very quick and easy cartoon and paid their US$10 for the exercise.

However, in my case, something was really going on – beyond my immediate comprehension. The attractive young artist was 'doing a real job' on me. Going into such detail that it gathered up a sizeable audience who were all talking excitedly amongst themselves and occasionally exchanging comments with the artist herself.

She was looking deeply into my eyes and I felt some form of magnetic attraction (perhaps I'd been away from home too long?)

At last she was finished, and we instinctively rose to our feet and then entered the warmest embrace imaginable.

The crowd cheered; I was confused. Everyone was chatting approvingly.

Helplessly I asked our interpreter, "What's next? Am I to be raffled? What's the girl's name, what am I supposed to do next?"

He quietly smiled and said, "No, it's not like that at all, this girl is respectably married but she has created considerable interest with her comments as she was sketching you".

He then spoke with her and explained to me the reason she took so long. "She saw the outside world through your eyes, and she was so excited; now she is sad because she will never know that world."

Artist – sketching the author

Artist

Sketch of author

Author, Roberto, Alex, Ronald – Moscow – September, 1990 (photo by Hannes)

Gran Canaria – Spain. The venue for this presentation.

With Dr Barbara Kolm of Austria, the creator of the Free Market Road Show
– The largest Libertarian event in the world

11

LIBERTY COULD BE GOOD FOR YOU TOO!

Liberty has been good for me.

I've been lucky enough to experience a few chance meetings that opened up my life to incredible opportunities, which I seized enthusiastically. This has opened unexpected doors in my life.

Although there have been many individuals whom I regard as teachers, who have made a vital difference to my life, there are four that stand out. These four individuals reinforced my belief that good ideas flow upward, rather than drift downward. These ideas often come from surprising sources.

The significant four

Any successful company I have run always followed a bottom-up management style and not a top-down dictatorship-style. I suspect that socialism is winning the battle of ideas today by focusing on promises to the individual. Our free marketeers, on the other hand, focus on fixing nations or the world and expecting individuals to believe there will be a few crumbs left over to 'trickle down' to them.

So, to succeed, let us focus on the individual. Absurd as it may seem, our free market movement has, in many ways, been focusing on the 'collective', i.e., the masses.

When comparing the 'individual' to 'society', let me refer to that wonderful challenge posed to us by Frédéric Bastiat, so many years ago:

> I affirm that political economy will have gained its objective and fulfilled its mission when it has conclusively proved this fact: What holds true for one man holds true for society. Man in a state of isolation is at once producer and consumer, inventor and entrepreneur, capitalist and worker; all the economic phenomena are performed in him, and he is, as it were, a society in miniature.[29]

Now let me mention the four outstanding individuals who had an impact on my life in different ways: Leonard E. Read, John Hospers, F.A. Hayek and Prince Philip.

Leonard E. Read

My world of the free market and the many interesting people I've met started when I was 16 years of age and living in Kalgoorlie, an Australian mining town. After high school, one afternoon, I was unpacking large engineering machinery crates from America in the back of my father's mining engineering business. As I pulled out the packing material, I noticed it had writing on it. It turned out to be pages of *The Freeman* magazine, published by the Foundation for Economic Education (FEE). These were all crumpled up. As you might imagine, this was many years before the advent of bubble wrap being used as packing material. I spread out these very crumpled magazine pages at home that night and was electrified to read the contents about free markets

---

[29] Frédéric Bastiat, *Harmonies of Political Economy*, 1850.

and individual responsibility. I can still remember the exhilaration of those new thoughts. Then, three years later while <u>the</u> editor of the *School of Mines Magazine*, I started publishing some of this material as an Australian version of these free market ideas. I was almost run out of town because Kalgoorlie was pretty much a union-controlled town at the time. The thought of being able to make it on your own without using union muscle to push the other guy around was not even to be considered.

I wasn't doing all that well, so I contacted the president of FEE, Leonard E. Read, and explained to him that their ideas were getting me into trouble. He politely wrote back explaining that these ideas really started with Aristotle and had been polished up over the years. He suggested that if I was having trouble defending myself against these antagonistic voices, it just meant that I had not spent enough time understanding my own position. With that in mind, he sent me some books, put me on his mailing list and that started our amazing relationship. Leonard E. Read was my mentor over many years. In 1982, some 30 years later, he invited me to FEE's office in Irvington, New York, to speak to his Board of Directors and tell them the story about the packing boxes. I told him that was a fairly ordinary story to be telling but he suggested that if I started the story, he would finish it off. He finished off by saying:

> Colleagues, our motto for the Foundation for Economic Education, is that ideas have consequences. We produced this small magazine and it went to the Timken Roller Bearing Manufacturing Company and then down through all the staff and ended up in the packing department. It was crunched up as packing material and sent off to the other side of the world where this young man smooths it out and goes on to form his own think tank, Mannkal Economic Education Foundation, based on the Foundation for Economic Education. Gentlemen, our ideas had more consequences than we had ever dreamt of.

This was the start of an adventure and I could spend many hours passing on some of the snippets of wisdom that Leonard E. Read passed onto me, over the years.[30]

By replying to my letters, Leonard E. Read opened-up my world. That is why I try very hard to reply to every letter or email that I receive.

Professor John Hospers

John Hospers and I had an unlikely but lasting friendship – he the philosopher and myself the prospector. My connection with John Hospers dates back to 1961 when I was still living in Kalgoorlie. I was then studying electrical engineering at the Kalgoorlie School of Mines. So, what has that got to do with philosophy? Let me tell you how I became interested in philosophy. Around 1960 Hugh Hefner launched *Playboy* magazine. My mother saw me with the magazine and suggested that I should not look at the pictures. So, being a totally obedient young boy, I only read the articles. Hefner's monthly editorial was called the Playboy Philosophy. If this was philosophy, well, I decided I would like to get more of it! So, I enrolled in an external philosophy unit with the University of Western Australia.

I enjoyed the course and was particularly captured by the textbook *An Introduction to Philosophical Analysis* by Professor John Hospers, then of the University of Minnesota. I liked the way Professor Hospers' mind worked so I started corresponding with him and we became lifelong friends. What led me to write my first letter to John in September 1961? My initial reason was entirely non-academic, as I just wished to share with him what I thought to be a very strange article that appeared in Australia's most un-academic weekly magazine called *Pix*. On that September day I was seated in a barber's shop reading a trashy magazine describing an 'unspeakably evil' textbook. Among other things, it proclaimed that "young students in Australian universities are being taught an evil

---

[30] Covered in more detail on pages 10-11 *Heroic Misadventures*.

philosophy from an evil textbook". I suddenly realised that this was exactly the same philosophy textbook that I had just completed, without detecting any sign of evil. What had I missed? That night, skimming through the book again (with the torn-out pages from the magazine), I realised that what the writer of that article had interpreted as 'evil' was what I admired about the book – namely, a sense of advocacy for heroic individual responsibility in the true Aristotelian sense.

After worrying about this misinterpretation for another 15 minutes, I wrote and asked Professor Hospers if I had missed something. Or was there, perhaps, another later edition that contained all this 'sinful' material? His response was measured and polite. He indicated that any academic who had a view on anything in particular was open to attack from time to time – often from surprising quarters. This magazine article was the subject of one of our later personal discussions and I suspect that he took this into account when he produced his 1972 book, *Human Conduct: Problems of Ethics,* where he expanded on the concept of personal responsibility in chapter 9 – 'Punishment & Responsibility'. John was forever focused on clear thinking and clear writing. The single point upon which the *Pix* article critic appeared to be fixated was their interpretation that John was promoting promiscuity. My interpretation was that John was advocating the virtues of personal responsibility.

Apart from thoroughly enjoying John's book during my philosophy study year, I often had occasion to refer to it throughout my life. I clearly remember one occasion when I referred to John's definition of 'beautiful'. This was during my time managing a hotel in Bali when I noted that the staff repeatedly burst into laughter every time an Australian guest described their food or a meal as 'beautiful'. The staff then explained to me that only Australians described food as 'beautiful' and, to them, the word was reserved for many other things but never for food. Hospers would have approved of the use of 'beautiful' in respect to food. As he said: "We are generally inclined to speak of objects as 'beautiful' when they arouse in us aesthetic experiences."

In 1997, at one of our meetings, John handed me a copy of the fourth edition of *Introduction to Philosophical Analysis*. It was clear that he had made many changes as he refined his own thoughts and, in my view, managed the Herculean task of editing his own work. The fourth edition totalled 282 pages in place of the third edition's 532 pages. There is one point of difference between the editions I can recall immediately. One of the new sections was *The Way the World Works: Scientific Knowledge*. I particularly enjoyed the subsection of that chapter titled 'A theory in geology'. Of geological theory he remarked:

> "We turn now to another theory, which includes data from astronomy and biology, with one science reinforcing the observational data of other sciences to form a coherent, unified theory."

This new section might explain the personal following that John developed among earth science professionals.

Little did I know in 1961 that, 12 years later, John Hospers would be the first Libertarian Party US presidential candidate or that I would become his lifelong friend right through to his death in 2011, at the age of 93. Fifty-seven years later, I'm exchanging views with his other fellow students, all over the world. We are grateful that he taught us how to think with clarity and develop our own personal philosophies – something which has assisted each of us over so many years. Each of our own John Hospers' experiences will be published in a John Hospers commemorative book. These fellow students are also working, in various ways, to republish updated versions of Hospers' books, including his classic 1971 volume *Libertarianism: A Political Philosophy for Tomorrow*, as well as making available his remarkable correspondence with the leading thinkers of his time, including Ayn Rand.

During the 50 years between my initial exposure to John Hospers and his death in 2011 we corresponded and enjoyed time together in several remarkable places. Let me mention just three occasions. The first was in 1984 when we invited John to London as the keynote speaker for the second Libertarian International Conference. A video of his talk is available at *Libertarianism.org/ media.* The second was in Los Angeles in the early 1980s. I was due to spend a day and night there on my way home to Australia from New York City, and John and I had scheduled a meeting. There was an airline complication that reduced my time in LA to four hours, confining me to the airport. I rang John from New York and explained that he would have to settle for me telephoning him from the LA airport and as I only had four questions for him, we could at least manage that. I clearly recall his response. "So, you will have four hours at the airport? Ron, you only have four questions for me, but I have 20 questions for you, so I will come to the airport and we can sit quietly for four hours." This we did while consuming several drinks. John's 20 questions were all focused on how I had come to the same conclusions on so many questions as he had, but from a completely different direction. He had been writing about the threats of unrestrained government and he saw me as someone who had been put out of business three times by my refusal to join the cosy cartels of occupational licensing. I'm sure he saw me as a laboratory rat with blood still pouring from my wounds, a worthy subject for his study!

The third was the last time we met. In was in 2004, when he spent the full day with my wife, Jenny, and me as our tour guide at the Ronald Reagan Presidential Library and Museum in California. For those who haven't had the pleasure of visiting it, let me mention that, in line with most US presidential libraries, it is financed entirely from private donations. The land, the buildings, the landscaping, even to Boeing's donation of Air Force One, meticulously reassembled in the garden, are all privately funded. The only exception is the underground vault which stores the official presidential papers. That is financed and securely managed by the US Federal Government. On that day, at that wonderful facility, John proved to be a magnificent tour guide. However, you can imagine my disappointment when we were refused entry into the official presidential documents under secure guard. The problem was solved when I called upon some John Hospers' clear thinking and loudly announced that John was a former presidential candidate. Well, a visit by three seemingly simple tourists was turned into something of a royal visit. From that point on we were appointed a special research guide and shown every imaginable document, including the handwritten speeches of Ronald Reagan, where he meticulously transformed run-of-the-mill speech writers' documents into very personally focused speeches, for which he was so noted. These are the documents that he was working on so late in the night when many Democrats were laughingly describing him as "dozing off in the Oval Office in the later hours". It was a remarkable day and John had so much to contribute to our depth of knowledge of the nature of the deep spirit of loyalty for which citizens of the United States are renowned.

Never one for pretence, John Hospers just wanted to be remembered as a teacher. His own words highlight his concerns for the criteria used by universities, even in those early days:

> I am mostly known as a writer of philosophy. But I always desired to be remembered as a great teacher. Universities, however, consider only a teacher's scholarly works and not their teaching ability. I want to be remembered as a philosophical instructor who could clarify questions and present good ideas clearly, avoiding vagueness and confusion in the presentation of ideas. That is probably my main legacy as a teacher and many of my students have come to remember me in just that way.

History has shown John Hospers to be a great teacher and I wonder if today's students will think as kindly about one or more of their own lecturers in 50 years' time.

It is difficult to conclude my memories of John Hospers in a manner that honours his own style, remembering also that he always liked to leave us with a challenge. These, the final words from John Hospers' Libertarian International Conference presentation in 1984, are a classic

John Hospers call to action:

There are now hundreds of books and articles demonstrating the superiority of the free-market, as well as books such as Ayn Rand's espousing their philosophy of liberty. Almost no such books existed a generation ago. A rising tide of Americans is now aware that government, not the market, is the cause of inflation, depression and poverty. These people, no longer children of Roosevelt's New Deal, are waiting in the wings, even in Washington, to reverse the course of the American economy, to remove the ball and chain of big government which still consumes the days and years of our lives. Even the academicians who have thus far turned to the government and defended it in return for favours to them, may come to realise that the Russian revolution which they have viewed so favourably is passé and that the real revolution, the revolution of 1776, of individual rights has taken place in their own land, unseen and unacknowledged by them.

The use of force by one government after another did not stop the clipper ships. In the end, they won the day and the wielders of governmental power had to go along or stagnate and die. In the same way, the soil of 1984, unlike the soil of say 1954, has been prepared for an outbreak of freedom which can pull even the welfare statists kicking and screaming into the 21st century and that is where we libertarians come in. We are the intellectual spearheads of the coming renaissance of liberty. Just as the intellectual influence of the Fabians propelled Britain into socialism a century ago, so the intellectual influence of libertarians can turn Britain, and indeed the world, back to individual liberty because now the soil has been prepared. The consequences of socialism in practise are increasingly plain for anyone with eyes to see. 'It's the essence of man,' said Aristotle, 'to make decisions'. His own decisions, not those made for him by others. To implement this simple but profound truth and to apply it over and over again, in its countless manifestations in our individual and social lives; that is our libertarian mission. Surely, it's the noblest of goals and I see no good reason why we should not be able to achieve it.

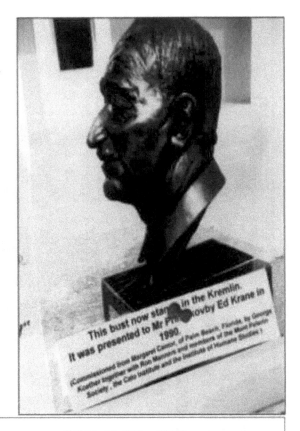

Bronze bust of F.A. Hayek that was presented to the Kremlin in Russia by Mr Ed Crane, President of the CATO Institute—September 1990

February 9th, 1977

Dear Mr. Manners, thank you for your kind letter of the 11th of last month which, because of my absence from here, has only just reached me. It was pleasant to hear from you after missin you in Australia.

I have so far seen little effect of my De Nationalization of Curren beyond a few kind but rather puzzlied reviews in newspapers and weekly journals. The people seem to find it difficult to believe that I meant it seriously.

Please note that from the end of this month my address will be
Urachstrasse 27
D-78 FREIBURG i.Brg.
West Germany.

Sincerely,

F.A.Hayek

The patient professor replied to my many questions

F.A. Hayek

Who could not be influenced by F.A. Hayek? Hayek's works can be described as a road map for the movement toward freedom and away from central planning. Back in 1975 my friend Ron Kitching phoned from the other side of Australia and asked me if I had heard Milton Friedman speak in Australia during his visit that year. We both agreed that Professor Friedman had injected a ray of light into Australia's economic debate. Kitching then said: "I think Australia is now ready for Hayek and I am expecting you to contribute some cash, along with Roger Randerson, Viv Forbes and a few other friends, so that we can issue an invitation and cover all expenses." Professor and Mrs Hayek visited in October 1976 and he gave a series of lectures as well as an ABC *Monday Conference* program (the *Q&A* of the 1970s).

I valued personal time spent with Professor Hayek in Hong Kong in 1978 and in Berlin in 1982 and despite his intellectual stature, he appeared to enjoy talking with 'mere mortals' like me. He said that we are closer to reality than many academics and I know that he sensed the importance of his ideas being expressed in language to which everyone could relate. Of course, this was well before the implosion of communism in the Soviet Union and Central Europe and the turn to market economies there, as well as in Latin America, Asia and even Sweden. All this transformation can be linked directly or indirectly to the work that Austrian-born Hayek had done during his long career spanning more than half a century. As he explained to me, he had spent much of his long life relentlessly developing and promoting the thesis that state control of economic life cannot enhance human wellbeing, it can only bring misery and poverty.

On other occasions, Professor Hayek reminded me of the important role that entrepreneurs should be playing in the battle of ideas. Once, I asked him to "slow down, I'm not an economist", to which he replied: "I'm glad you are not an economist. We economists simply dream our ideas and think our thoughts, but you go out and fire the bullets!" I walked away from that meeting feeling that I had been given a useful role in society. On another occasion he explained to me why, quite often, I would find that academics disliked business people. He said that the big difference between business people and academics was that business people enjoyed volatility as they see it as opportunity, while academics craved certainty, such as tenure. He mentioned that some academics felt it to be unjust that some business people were paid more than academics and that it drove them crazy to see business people happier than many academics. All these snippets of wisdom were passed on to me with Hayek's typical tongue-in-cheek whimsical style even though personally I have not experienced any conflict between academics and entrepreneurs. Again, I value my personal correspondence with Professor Hayek.

HRH Prince Philip – Duke of Edinburgh

Prince Philip, the fourth individual on my list, was well known to the other three. In fact, FEE.org featured an excellent article on Prince Philip as an individualist (A Prince Replies to Machiavelli: Prince of England on the erosion of freedom, 01 February 1978.) In 1956 Prince Philip, a keen observer of industrialisation and its effect on individuals, realised that the three main community sectors – industry, trade unions and government – were not talking to each other. He devised a plan to select 100 potential leaders from each of these three sectors and 'lock them up together' for three weeks. This meant they would be living together, travelling together and learning together. He felt that lifelong bonds would be forged between warring parties and the benefits would become obvious during subsequent years. These conferences, held every four years, have generated over 2,500 well-connected individuals, still vigorously talking and learning from each other. It was my honour to be selected for the

1968 intake and that gave me the first of five opportunities to share experiences with Prince Philip over the years.

In 1968, Prince Philip was ahead of his time. Many of his words still ringing true today. He said:

> Ideas are coming into Australia from the young people and unfortunately there is a time delay before they permeate through to the old. Don't leave the change too long. Be tolerant but not permissive with our young. They are as much the children of their age as we were of ours.

He taught us how to ask questions by reminding us that, the first time we ask anyone a question, we will only receive a polite answer. This is because they are unsure if we really want to know. The second time we ask that question they will take us slightly more seriously and again give a partial answer. It's only the third time we ask the same question, still being polite, that we will really get inside their mind. Once they realise how serious we are, they will open up and give us the true story. Prince Philip said: "That's the answer I want you to bring back to me, fully refined and fully focused – and don't waffle" He recognised that a single approach didn't suit everybody. "We can bring our children up by the book as long as we use a different book for each child," he said. He asked us to think and speak as individuals and not just be a spokesperson for any organisation or government. He told us to get over our great Australian distrust of excellence. These were the two points that he wanted to leave us with. Firstly, that we should come to our own conclusions and act as an individual to avoid what is now termed group-thinking. He was so focused on individualism that when he invited us to Buckingham Palace for the 50th Anniversary Reunion of the Commonwealth Study Conference, he said: "… and you can't bring your wives or partners because I'm not bringing mine."

The second point was to remember to ask questions that give us correct answers.

These comments, and the 1968 study tour itself, inspired me to set up Mannkal Economic Education Foundation, some 30 years later. Prince Philip's personal training, for me, was a profound gift and the task of extending this profound gift lies with us who have been the beneficiaries.

Mannkal Economic Education Foundation

Mannkal Economic Education Foundation's program is a combination of the influence of all four of those individuals who I have mentioned. The first three influences – Leonard E. Read, John Hospers and F.A. Hayek – were heavy on philosophy but very light on strategy. The Prince Philip input, to me, was completely devoid of philosophy with a 100 per cent focus on strategy. To use a mining metaphor, what we at Mannkal have done is blend philosophy and strategy, run them down through a crushing plant and screened them to a size that just suits the task in hand. This is a photo of our finished product. This group of students were attending the 2017 Friedman Conference in Brisbane, Australia.

After putting these thoughts and actions through our corporate 'crushing, screening and refining process', we are increasing our output of smart, questioning and useful young Australians. Over 1,500 so far. Young people are interviewed and selected for events that will expose them, many for the first time, to economic and political philosophical principles that promote the virtues of individual responsibility. They discover that this is more difficult than the easier alternative of living off the efforts of unsuspecting taxpayers, many of whom are less well-off than the recipients of handouts. This leads these young people to study the often unintended long-term consequences of many of today's short-term legislative solutions and policy proposals.

As Leonard E. Read said, ideas have consequences, particularly when applied to the study of liberty. Liberty is a simple idea, but it's also the linchpin of a complex system of values and practices: justice, prosperity, responsibility, toleration, cooperation and peace. Many people believe that liberty is the core political value of modern civilisation itself, the one that gives substance and form to all the other values of social life.

In 2018, Mannkal Foundation's 21st Anniversary year, the momentum is building to the point where it is taking me away from my life-long involvement in mining and management and I enjoy measuring the results of our labours.

Why I'm an optimist

People ask me: "How can I remain an optimist when all around us we see deterioration in the standards by which we measure excellence? In business, in family, in education and particularly in politics."

Politics is almost universally broken, nearly beyond repair (perhaps with the exceptions of Switzerland and Estonia). In many countries it is completely broken.

Here is why I'm an optimist.

Leonard E. Read taught me that life is not a numbers game – you don't need the biggest gang to achieve your goals. History is full of examples of how small groups of individuals have achieved amazing results.

Leonard used the example of Christianity. He said, "Jesus only had a bunch of 12 in his team and one of them even turned out to be a treacherous bum." More recently, in the late 1980s, three people with advice from a fourth, called the bluff on communism and communism collapsed.

That was achieved by a few telephone calls to the Pope from President Ronald Reagan and Prime Minister Margaret Thatcher. The calls could have gone like this:

"Good morning, Your Holiness. Our friend and economic advisor, Friedrich Hayek, tells us that Russian superiority is a myth and it will topple if we expose them for what they really are, just a bunch of frauds. Let's pull the rug from under them and see what really happens."

Now, that's exactly what happened, and Russia fell over, without a single life being lost too!

Those of us who were fortunate enough to have participated in the CATO Institute's Transition to Freedom Conference, in Russia, in September 1990, would have experienced first-hand this Russian 'myth of excellence' (explained in detail in an earlier chapter).

Russia's economy is shrinking whilst most other economies are growing and will continue to grow as long as we effectively promote the benefits of free, transparent trade with each other.

The strategic task of organisations such as the Mont Pelerin Society and the Atlas Network of intricately related think tanks, around the world, if focused with laser-like precision, will concentrate the wisdom of our free market message into effective bullets for us to fire. We can win this ongoing war of ideas.

I know that in 1947 the original founding members of the Mont Pelerin Society called themselves the 'Remnant'. Following those darkest days of World War II, it looked as though government central planning would remain forever. They set themselves the task of assembling the essential guidelines for a free market economy, in the hope that future generations would reach out for the ideas in an attempt to throw off the chains of central planning.

Their idea was simply to store all this wisdom in an imaginary intellectual time-capsule that could be opened by a desperate future generation. Fellow travellers – that future generation is us. We should waste no time in opening up that time-capsule.

These ideas have been further refined by subsequent intellectual giants, which now gives us an effective narrative with which to work.

We now have the ammunition so the focus must be on strategic execution, rather than simply maintaining our 'Remnant' role. We may have fed the troops from the comfort of our ivory towers, but it is now time to move to the front line.

We need to experience the exhilaration of firing the intellectual bullets and decisively winning by focusing on what we can offer to individuals, rather than the masses, by developing 'bottom up' strategies.

I believe we have the right people to put this strategy together and see no reason for us not to win this crucial battle of ideas. I'm an optimist because there is a growing appreciation of the role of the free market entrepreneur. These entrepreneurs are impatient and have an essential part to play in overcoming intellectual inertia.

Academia must learn to love these free market entrepreneurs because without them we will be forever stuck in the time-capsule mode. The stakes are high and because of the quality of our ideas, all we need to do is lift ourselves up from being armchair observers and graduate into frontline activists.

Friends let the battle commence.

A Prince Replies to Machiavelli: Philip of England on the Erosion of Freedom

Wednesday, February 01, 1978

E.G. West

Philip:

Eir:

SANDRINGHAM, NORFOLK

28 February 2019

Dear Ron Manne + Paul Meyer,

Many thanks for your letter. I am delighted to know that you even remembered the 'Study Conference'! We must have hit on a successful formula, since the experience seems to have helped a good many people. They still continue, although in a somewhat modified formula, under the aegis of my daughter Anne.

I am very pleased to know that you think the conference may have helped you, although I suspect you would have done pretty well anyway!

Yours sincerely

Philip

The Mont Pelerin Society

Sir Karl Popper[31], the pre-eminent philosopher of science of recent times, described the Mont Pelerin Society, in a speech to the Annual Meeting of the American Economic Association, New Orleans, January 4 1992;

> … "But Hayek did not confine himself to writing these politically so amazingly powerful works. Although a great scholar and distinguished gentleman, rather reserved in his way of living, thinking and teaching, and averse to taking political action, he founded, shortly after the Second World War, the Mont Pelerin Society. Its function was to provide a balance to the countless intellectuals who opted for socialism. Hayek felt that more had to be done than writing papers and books. So, he founded a society of scholars and practical economists who were opposed to the fashionable socialist trend of the majority of intellectuals who believed in a socialist future. The society was founded in Switzerland in 1947 on Mont Pelerin, on the southern shores of Lake Geneva. I had the honour of being invited by Hayek to be one of the founder-members…

> …Its first and perhaps greatest achievement was, I feel, to encourage those who were fighting the then overwhelming authority of John Maynard Keynes and his school…

> …I could experience the growing undermining by leftist teaching which, in the first few years after the war, had been immensely powerful…

> …It was Mises who advanced the first and fundamental modern criticism of socialism: that modern industry is based on a free-market, and that socialism, and especially "social planning", was incompatible with a free-market economy and consequently bound to fail."…

> ….However, not being an economist myself, I am probably not competent to assess the historical influence of the Mont Pelerin Society. This is a task—I think an important task—for future historians of economic doctrines and economic policies."

---

[31]   Sir Karl Popper's long correspondence with F.A. Hayek appears as Vol. XIX of the *Collected Works of F.A. Hayek.*

APPENDIX 1

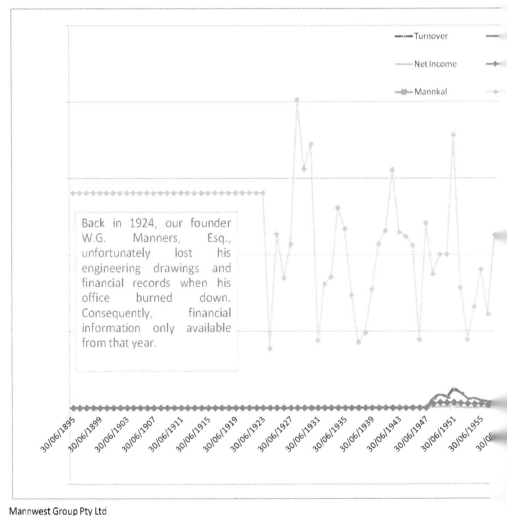

Legend: Turnover, Net Income, Mannkal

Text box within chart:
Back in 1924, our founder W.G. Manners, Esq., unfortunately lost his engineering drawings and financial records when his office burned down. Consequently, financial information only available from that year.

X-axis labels: 30/06/1895, 30/06/1899, 30/06/1903, 30/06/1907, 30/06/1911, 30/06/1915, 30/06/1919, 30/06/1923, 30/06/1927, 30/06/1931, 30/06/1935, 30/06/1939, 30/06/1943, 30/06/1947, 30/06/1951, 30/06/1955, 30/06/...

Mannwest Group Pty Ltd
Hayek on Hood,
3/31 Hood Street,
Subiaco, Western Australia 6008
www.mannwest.com

120 Years of Service
1895 - 2015

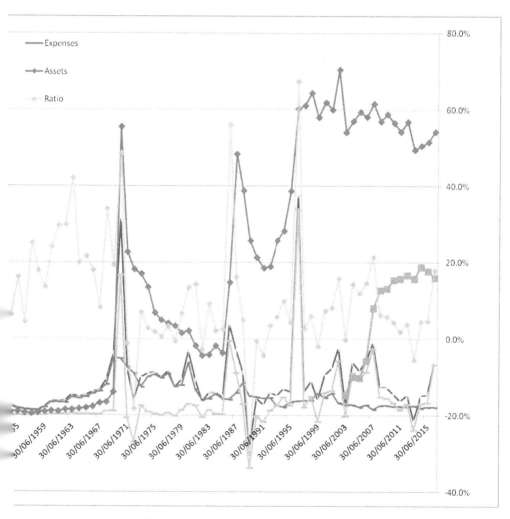

Special Report to Commemorate our
120th Anniversary

BIBLIOGRAPHY

Bastiat, Frédéric, 1996 (1850), *Harmonies of Political Economy*, (Foundation for Economic Education) New York.

Bastiat, Frédéric, 1998 (1850), *The Law*, (Foundation for Economic Education) New York.

Boetke, Peter, J., 1993, *Why Perestroika Failed*, (Routledge) New York.

Brown, Harry, 1971, *The Complete Guide to Swiss Banks*, (McGraw Hill) New York.

De Pree, Max, 1992, *Leadership Jazz*, (Doubleday Business) New York.

Fishall, R.T., 1981, *Bureaucrats: How to Annoy Them*, (Sidgwick & Jackson) London.

Hayek, F.A., 1944, *The Road to Serfdom*, (Routledge Press) Abingdon-on-Thames, Oxfordshire.

Hospers, John, 1971, *Libertarianism: a Political Philosophy for Tomorrow*, (Nash Publishing) Ann Abor, MI.

Hospers, John, 1995, *Human Conduct: Problems of Ethics*, (Cengage Learning) Boston, MA.

Hospers, John, 1997, *An Introduction to Philosophical Analysis*, (Prentice Hall) Charlottesville, VA.

McDonald, Geoff, 1982, *Red Over Black*, (Veritas Publishing Company) Unley, SA

Mises, Ludwig von, 1981, *Socialism: an economic and sociological analysis*, (Liberty Fund) Indianapolis, IN.

Peters, Tom, 1994, *Liberation Management*, (Fawcett Columbine) New York.

Samuelson, P.A., 1969, *The Way of an Economist*, (MacMillan) London.

Solzhenitsyn, Aleksandr, I., 1973, *The Gulag Archipelago*, (Harper & Row) New York.

Windschuttle, Keith, 2002, *The Fabrication of Aboriginal History, Volume one, Van Dieman's Land 1803 – 1847*, (Macleay Press) Paddington, NSW.

INDEX

Aristotle 47, 221, 227

Arkaroola 26, 27, 35, 239

Atlas Economic Research Foundation / Atlas Network 34, 234

Australian Constitution 49, 131, 152, 155

Australian Flag 49

Australian Tax Office (ATO) 15, 38, 40, 52, 117, 175, 198

Balzano, James 4

Bank National de Paris (BNP) 201

Barnett, Colin 189, 195

Barton, Peter 44

Bastiat, Frédéric 102, 134, 135, 220

Bennett, Mark 185

Boetke, Peter J. 212

Boudreaux, Don 9

Bradford, Kerry 192

Brand, David 176

Bre-X Minerals Ltd 183, 204, 205

Bremer Range 173

Browne, Harry 142

Buchanan, James 45, 187, 210

Campbell, Joseph 207

Cappello, Anthony 11, 15

Carroll, Judy 11

Casey, Doug 193

Cash, Sam 168

CATO Institute 22, 45, 209, 214 234

Central Norseman Gold Corporation 202

Cicero 153

Clark, Colin 110

Clarke, Ron 56, 57, 110

Clyde, James (Lord) 133

Clyne, Peter 120, 143

Communism 3, 34, 157, 211, 229, 234

Compton, George 4, 172

Confucius 40

Connor, Rex 188

Coombs, T. C. 181

Cox Brothers 169

Crane, Ed 210, 213

Croesus Mining 38, 42, 43, 54, 179, 180, 182-185, 192, 196, 197, 199-207

Curtin University 186, 192

De Pree, Max 42, 43

Deinega, Alex 215

Delta Gold Ltd 179

Donaldson, Geoff 191

Duncan, Travers 180

Economics of Politics 45, 46,

Eldorado Gold 54, 183, 203-205

Elkington, Dick 172

Elkington, John Henry 172

Elkington, Norma 172

Emerson, Ralph Waldo 47

Enterprise Zones 60, 62, 63

Eric Andersons 169

Eshuys, Ed 173, 175

Esperance 29, 44, 45, 51, 153, 168

Estonia 24, 152, 213, 233

Fadden, Sir Arthur 54

Fagan, Bob 192

Fisher, Antony 91

Flat tax rate 151

Forbes, Viv 229

Foroux, Darius 43, 44

Foundation for Economic Education 29, 31, 220, 221

Fraser, Malcolm 84, 191

Friedman, David 152

Friedman, Milton 105, 109, 113, 156, 229

Friedman, Rose 105

Gencor/Billiton 204

Gilbertson, Brian 204

Gilder, George 213

Gissurarson, Hannes H. 215

Golden Mile 37

Goode, Keith 185

Great Central Mines NL 54, 179

Griffin, Tim 193

Grill, Julian 64

H. G. Palmer 169,

Hains, David 39, 55

Hannan South Gold Operations 183

Hannan's Club 178,

Harris, Ralph (Lord) 91

Hartrey, Tom 174

Hayek, F. A 14, 42, 70, 135, 220, 229, 230, 232, 234, 236

Hazlitt, Henry 3, 116

Hefner, Hugh 222

Henry, Patrick 112

Heroic Misadventures 4, 9, 13, 15, 17, 33, 38, 52, 165, 176,180, 190, 222

Hobson, R. A (Hobby) 167

Hong Kong 9, 10, 61, 62, 65, 68, 70, 71, 75-89

Hospers, John 220, 222-224, 226, 232

Illarionov, Andrei 213, 215

Institute for Economic Affairs 91

Jaffe, Dennis 49

Japan 52, 115, 211

Jefferson, Thomas 29, 41, 67, 72, 98, 112, 137

Kalgoorlie 4, 9, 16, 28, 30-33, 37, 38, 50, 51, 62, 63, 70, 167, 168, 170,-172, 174-176, 178, 182, 184, 197, 199-202, 204-206, 220- 222

Kalgoorlie Chamber of Commerce 70

Kalgoorlie School of Mines 31-33, 37, 38, 51, 167, 176, 200, 221, 222

Kalmin Exploration Ltd 52, 174, 175

Kambalda 62, 171, 176-179, 201

Keynes, John Maynard 236

Kiruna 58, 179

Kitching, Ron 212, 215, 229

Kitson, H. M (Harry) 54, 55, 182, 184

Laffer, Arthur B. 156, 157

Lalor, Chris 180

Lee, Simon 55

Leon, Roberto Salinas 215

Lexus 73

Lord Acton 14

Lydia (Western Turkey) 199

MacKinnon, Barry 64

Maggie Hays Hill 173

Malga Minerals NL 52

Mannelksploration Pty Ltd 172-174

Manners, Charles 4

Manners, Nancy 4

Manners, W. G. 7, 9, 26, 30, 50, 117, 177, 215

Mannkal Economic Education Foundation 21, 34, 54, 221, 232, 233

Mannkal Mining Pty Ltd 51, 54, 185, 203

Mannwest 34, 49, 52, 54, 55, 183, 188

Marx, Karl 3, 104

Maxwell, Clyde R. 55, 120, 122, 199

Maxwell, Philip 192

May & Mellor 181

McIntyre, Colin 192

Mencken, H. L. 117

Menzies Credit Squeeze 51, 166, 168

Metals Exploration 171

Metana Minerals 179

Mill, John Stuart 94

Mines and Money Conference 65

Mont Pelerin Society 15, 34, 146, 234, 237

Moore, Patrick 92

Morgan, W. M. 178

Moscow 187, 209, 212, 213, 215

Mount Monger 180

Native Title Act 70, 182, 183, 190, 202

New York City 129, 170, 225

Nickel 9, 30, 51, 52, 63, 166, 169,- 172, 175-177, 179, 198, 201

Nixon, Richard 179

Normandy 179

Norseman 22, 51, 168, 174, 180, 184, 192, 202, 206, 207

Northern Development and Economic Vision (ANDEV) 67

Nozick, Robert 187

Nugan Hand 52, 175

O'Brien, Avens 3

Oliver, John 167, 176, 182, 200

Palmer, Tom 212

Parisotto, Libero 192

Parry, Keith 55, 167, 180

Payne, Thomas 112

Peters, Tom 42

Peterson, Jordan 19

Philip, Prince – Duke of Edinburgh 33, 220, 231, 232

Pincus, Jonathan 157

Plimer, Ian 26

Popper, Karl 237

Poseidon 171, 173, 175, 178, 191

Public Choice Theory 45, 46, 51, 150, 187, 210

Puccinellia 44, 45, 153

Putin, Vladimir 213, 214

CRA (Rio Tinto) 181, 182, 200

Raico, Ralph 213

Rand, Ayn 19, 92, 224, 227

Randerson, Roger 229

Ray Porter & Partners 181

Read, Leonard E. 10, 31, 89, 147, 220-222, 232-234

Reagan Administration 52, 175

Reagan, Ronald 63, 225, 226, 234

Regressive taxation 104

Reid Murray 169

Rinehart, Gina 189, 195

Rio Tinto (CRA) 181, 200, 201

Risstrom, Eric 99

RMIT 23

Roberts, Paul Craig 209

Rodgers-Falk, Josh 2

Rothbard, Murray 11

Ruddock, Ian 193

Russia 22, 34, 45, 209, 211, 213, 214, 234

Russian toilet paper 211

Samuelson, Paul 185, 186

Seldon, Arthur 91, 94

Shames, Laurence 44

Shierlaw, Norm 171, 178, 191

Siberia 175

Silver Lake Shaft 177

Smith, Adam 96, 98, 132, 133, 135

Socialism 3, 34, 88, 97, 102, 104, 111, 135, 146, 147, 158, 198, 199, 219, 227, 236

Solzhenitsyn, Aleksandr I.157

Sons of Gwalia 179, 180

Soviet Academy of Sciences 209

Spain 15, 214

St. Petersburg 209, 213, 215

Stacey, Bill 10

Subaru 52

Sweden 48, 52, 58, 61, 62, 179, 229

Sykes, Trevor 178

Tangney, Dorothy 45

Texas 214, 215

Thales 91

The Fabrication of Aboriginal History 190

Theseus Exploration NL 173

Thomas, Gavin 191

Tier, Mark 68, 82

Timken Roller Bearing Manufacturing Co. 221

Tulloch, Gordon 45

Turkey 44, 153, 199, 204

Twain, Mark 7, 16, 100

Tzu, Lao 40, 41

Ulyatt, Chris 11

Union Miniére 173, 174, 185

University of Notre Dame 46

Van Groenwoegel, H. J. 146

Volvo 52

Von Mises, Ludwig 21, 99

Von Szent-Gyorgyi, Albert 207

Washington, George 112

Weasel words 186

Wesfarmers 169

Western Civilization 3, 13, 20, 34

Western Mining Corporation 51, 61, 169, 171, 176, 177, 180, 201, 202

Westpac Bank 181, 182, 200

Westralian Nickel NL 175, 198

Whitcomb, Nan 242

Whitlam Gough 84, 85, 188

Whyte, Rod 201

Williams, Rohan 192

Windarra 175

Windschuttle, Keith 190

Woodside Petroleum 191

Wright, Nicola 11

Wylie, W. R. A (Bill) 75

Yew, Lee Kuan 191

Lightning Source UK Ltd.
Milton Keynes UK
UKHW022021130220
358691UK00005B/183